Fashion and Technology

Fashion and Technology
A Guide to Materials and Applications

ANETA GENOVA AND KATHERINE MORIWAKI

PARSONS SCHOOL OF DESIGN

Fairchild Books
An imprint of Bloomsbury Publishing Inc

BLOOMSBURY
NEW YORK · LONDON · OXFORD · NEW DELHI · SYDNEY

Fairchild Books
An imprint of Bloomsbury Publishing Inc

1385 Broadway
New York
NY 10018
USA

50 Bedford Square
London
WC1B 3DP
UK

www.bloomsbury.com

**FAIRCHILD BOOKS, BLOOMSBURY and the Diana logo
are trademarks of Bloomsbury Publishing Plc**

© Bloomsbury Publishing Inc, 2016

All rights reserved. No part of this publication may be reproduced or transmitted in any form or by any means, electronic or mechanical, including photocopying, recording, or any information storage or retrieval system, without prior permission in writing from the publishers.

No responsibility for loss caused to any individual or organization acting on or refraining from action as a result of the material in this publication can be accepted by Bloomsbury Publishing Inc or the author.

Library of Congress Cataloging-in-Publication Data

Genova, Aneta, author.
Fashion and technology : a guide to materials and applications / Aneta Genova and Katherine Moriwaki, Parsons School of Design.
pages cm
Includes bibliographical references and index.
ISBN 978-1-5013-0508-5
1. Clothing and dress—Materials. 2. Fashion design—Technological innovations. 3. Electronic textiles. 4. Wearable technology. I. Moriwaki, Katherine, author. II. Title.
TT557.G46 2016
746.9'20285--dc23
2015027448

ISBN: PB: 978-1-5013-0508-5
ePDF: 978-1-5013-0509-2

Typeset by Lachina
Cover Art Credit: Stephanie McNiel
Cover Design: Eleanor Rose
Printed and bound in China

CONTENTS

Extended Table of Contents **vii**
Preface **xi**
Acknowledgements **xv**

1. The Integration of Technology into Fashion **3**
2. How to Design with Electronics **29**
3. How to Design with Conductive and Reactive Materials **85**
4. How to Design with Existing DIY Kits **121**
5. Introduction to Digital Fabrication **155**
6. Introduction to Code **193**

Glossary **219**
Resources **223**
Credits **000**
Index **000**

EXTENDED TABLE OF CONTENTS

Preface **xi**

Acknowledgements **xv**

1 The Integration of Technology into Fashion 3

Materials 5
Process 7
Distribution 8
Technology on the Body 8
Wearable Computers 10
Computational and Electronic Fashion 11
DIY and the Maker Movement 13
Makers in Action: Popularizing Tools 14
Design Aesthetics 15
   CASE STUDY: VEGA EDGE 17
     INTERVIEW: KATE HARTMAN (COLLABORATOR, VEGA EDGE) 20
     INTERVIEW: JOANNA BERZOWSKA 23
For Review 27
For Discussion 27
Key Terms 27

2 How to Design with Electronics 29

Electricity 30
Electric Circuit 31
   Series vs. Parallel 32
   Flexible Circuits and Electronics 34
Basic Components of a Circuit 36
   Batteries 36
   Resistors 37
     Photoresistors 38
     Potentiometers 38
   LEDs 39
   Switches 42
   CASE STUDY: FLIP SWITCH 45
   INTERVIEW: KATHARINA BREDIES (CO-CREATOR OF THE FLIP SWITCH) 49
Tutorials 51
   TUTORIAL 1: BASIC SOFT CIRCUIT WITH VISIBLE AND INVISIBLE STITCH LINE 51
   TUTORIAL 2: SOFT CIRCUIT WITH LEDs CONNECTED IN PARALLEL VS. SERIES 56

TUTORIAL 3: MOMENTARY SWITCH (PLASTIC ZIPPER) AND MAINTAINED SWITCH (METAL ZIPPER) CIRCUITS FOR A SHORT ZIPPER 61

TUTORIAL 4: MOMENTARY SWITCH (PLASTIC ZIPPER) AND MAINTAINED SWITCH (METAL ZIPPER) CIRCUITS FOR A LONG ZIPPER 69

INTERVIEW: KOBAKANT 77

For Review 81
For Discussion 81
Books for Further Reading 81
Online References 81
Key Terms 82

3 How to Design with Conductive and Reactive Materials 85

Conductive Materials 86
    Conductive Thread and Yarn 86
    Conductive Wool 89
    Conductive Fabric 89
    Conductive Loop Fastener 89
    Conductive Tape 91
    Conductive Paints and Inks 91
Metal Fasteners and Closures 92
    Hook and Eye 92
    Sew-On Snaps 93
    Zippers 93
Reactive Materials 94
    Thermochromic Inks 94
    Wind Reactive Inks 95
    Photochromic Materials 95
        UV Reactive Paints and Inks 95
        UV Color Changing Thread 95
        Plastic Color Changing Resin Concentrates 95
    Hydrochromic Inks 95
Electroluminescence 101
    TUTORIAL 1: SWITCH TUTORIAL—CONDUCTIVE HOOK AND EYE CLOSURES 103

TUTORIAL 2: SWITCH TUTORIAL—METAL SNAPS AS CLOSURES 109

INTERVIEW: ISABEL LIZARDI (FROM BARE CONDUCTIVE) 115

For Review 118
For Discussion 118
Books for Further Reading 118
Online References 118
Key Terms 119

4 How to Design with Existing DIY Kits 121

Overview of DIY Electronics 122
    An Expanded Toolbox 123
    Tools 123
    Resources 127
    DIY Electronics 128
Anatomy of an E-Textile Toolkit 129
    Peripherals 133
        Sewable LEDs 133
        Battery Holders 134
        Sensors 135
        Other Peripherals 136
Insight into the Collaborative Process between Designers and Technologists 136
    CASE STUDY: CLIMATE DRESS BY DIFFUS STUDIO 137
    INTERVIEW: MICHEL GUGLIELMI (CO-FOUNDER OF DIFFUS DESIGN GROUP) 140
    TUTORIAL: HOW TO USE A DIGITAL MULTIMETER CONTINUITY/CONNECTIVITY TEST 143
    INTERVIEW: BECKY STERN (DIRECTOR OF WEARABLE ELECTRONICS AT ADAFRUIT INDUSTRIES) 148

For Review 151
For Discussion 151
Books for Further Reading 151
Key Terms 152

5 Introduction to Digital Fabrication 155

Overview of Digital Fabrication 156
Types of Fabrication Machines 158
    Laser Cutting 158
    CNC Router 159
    3D Printing 162
Collaborative Process 163
    CASE STUDY: BRADLEY ROTHENBERG
       FOR THREEASFOUR 165
Future Trends 169
    CASE STUDY: TRIPTYCH BY TANIA
       URSOMARZO 170
    TUTORIAL 1: LASER CUTTER 174
    TUTORIAL 2: 3D MODELING / 3D PRINTING 177
    INTERVIEW: GABI ASFOUR (FROM
       threeASFOUR) 183
    INTERVIEW: BRADLEY ROTHENBERG 187
For Review 190
For Discussion 190
Books for Further Reading 190
Key Terms 191

6 Introduction to Code 193

Code 194
Creative Coding 195
Using Code 195
    Patterns for Print and Surface
       Treatments 196
    Shape and Form 198
Programming Concepts 200
Processing 200
    TUTORIAL: LEARNING CODE 202
    CASE STUDY: FASHION FIANCHETTOS 208
    INTERVIEW: OTTO VON BUSCH 212
    INTERVIEW: CAIT REAS 216
For Review 218
For Discussion 218
Books for Further Reading 218
Key Terms 218

Glossary 219
Resources 223
Credits 000
Index 000

PREFACE

In 1801 Joseph Marie Jacquard demonstrated the Jacquard weaving loom for the first time. The machine, which simplified the process of weaving textiles such as brocade, damask, and matelassé, was controlled with punch cards, with each row of weaving represented by one card. This early machine was one of the inspirations for early computing, which also relied on punch cards to "program" the sequence of operations the computer executed. That the origins of computing should have roots in the world of fashion is somewhat fitting in light of this book.

This textbook elaborates on a particular relationship fashion has with computational technologies. Though fashion and technology have long intersected, recent developments in computation have enabled the adoption of new methods, tools and approaches to design. Communities of practice outside of fashion have been enabling the creation of garments and accessories that take advantage of emerging technologies, and while fashion has also kept apace of recent developments, there really isn't adequate documentation of this work from an instructional perspective, particularly with a focus on fashion design students.

We are working from a framework that connects the world of electronics and code, where physical objects become enmeshed with emergent capabilities that would not be possible without computation. Computers have become such an integral part of contemporary life that there are several "computer interfaces" which no longer appear to be such. We don't question their presence in our lives as we do the latest gadget; they are, as theorist Wendy Chun argues, a part of our habitat. There are obviously other technological trends which impact fashion (for example, current work on biologically engineered design and the use of living organisms to grow and cultivate new materials) but computational design processes involving electronics, digital fabrication, and code have established a design space that is rapidly becoming a part of the mainstream design practice. It is in this context that we discuss technology, not as a distanced conceptual enterprise, but as part of grounded practical techniques and approaches that designers can incorporate now.

Our intent is to introduce the reader to this interdisciplinary new area of practice, mapping a design space that is rapidly expanding—both in innovation and applicability to the field. Our primary audience is the college level fashion design student, but this book is of interest to anyone who is curious about how technology integrates into fashion. As with any subject that is dynamically changing, it is difficult to capture every recent development. This book is not meant to be an exhaustive catalogue of every material, technique, and tool currently available. Instead, we are aiming to identify broad categories and trends in a way that orients the reader towards new methods and approaches that may be incorporated into design practice. It is with great excitement that we present this book to our readers and look forward to what are sure to be exciting developments in this rapidly changing, fast paced, emerging area of design.

Coverage and Organization

This book is organized into six chapters. The first chapter is an overview of the intersection of fashion and computational technology with an emphasis on understanding how working within varied communities of practice has brought us to the present day. The remaining chapters cover practical aspects of working with technology. Chapter 2 focuses on basic electronics. Chapter 3 focuses on an overview of reactive materials. Chapter 4 orients the reader to the DIY electronics community, which has been partly responsible for the popularization of wearable technology and fashion. We then proceed to Chapter 5, which covers digital fabrication. Finally, Chapter 6 provides an overview of programming.

Features

This book is intended to support students of design who would like to familiarize themselves with the various ways in which they can consider adopting computational technologies into their work, whether this is through electronics, digital fabrication, or code. Our book focuses on process and orients the reader to current possibilities. Rather than attempt to have fashion designers become engineers or programmers, we introduce the reader to the design opportunities presented by each area, provide working definitions and vocabulary, identify current communities of practice that may be suitable for collaboration, and provide simple tutorials as a starting point for individual learning. This book is meant as a practical guide, and as the technological needs of fashion change, there will be a need for designers who understand this space and can serve as visionary leaders and informed collaborators.

Each chapter contains learning features to support these goals:

- **Case Studies** profile projects relevant to each chapter, providing exemplary work that embodies each chapter theme. Each Case Study draws from the work of contemporary designers in the field.
- **Interviews** present conversations with designers working at the intersection of fashion and technology. The interviewees are diverse in their career path and training, and provide broad insight into the field and their design process.
- **Tutorials** outline hands-on, step-by-step instructions for how to use the technologies and methods described in the book. These activities are designed as a starting point for individual learning and experimentation.

Fashion and Technology STUDIO:

Student Resources

This book also features an online multimedia resource—*Fashion and Technology STUDIO*. The online *STUDIO* is specially developed to complement this book with rich media ancillaries that students can adapt to their visual learning styles to better master concepts and improve grades. Within the *STUDIO*, students will be able to:

- **Study smarter** with self-quizzes featuring scored results and personalized study tips.
- **Review concepts** with flashcards of essential vocabulary.
- **Watch videos** with additional tutorials and bonus interview footage.

STUDIO access cards are offered free with new book purchases and also sold separately through Bloomsbury Fashion Central (www.BloomsburyFashionCentral.com).

Instructor Resources

- **Instructor's Guide** provides suggestions for planning the course and using the text in the classroom, with supplemental assignments and lecture notes.
- **Test Bank** includes sample test questions for each chapter.
- **PowerPoint® presentations** include images from the book and provide a framework for lecture and discussion.

Instructor's Resources may be accessed through Bloomsbury Fashion Central (www.BloomsburyFashionCentral.com).

ACKNOWLEDGEMENTS

We would like to thank Parsons School of Design and The New School University for providing such an inspiring place to work. It is a privilege to have such great colleagues in the School of Fashion, the School of Art, Media and Technology, and the MFA Design + Technology program. We share deep gratitude for the support of Sven Travis who over the years has supported us with generosity and openness to the collaborative work we've pursued. We thank Anne Gaines, Dean of the School of Art, Media and Technology, for her support of our writing and teaching. Many acknowledgements of appreciation to Fiona Dieffenbacher and Francesca Sammaritano, who provided feedback on the book when still in its proposal stage and beyond. We are grateful for Fiona's continued support as Director of BFA Fashion. And of course we are deeply indebted to the many students we have worked with over the years for bringing energy, enthusiasm, and inspiration to everything they do.

We thank our many contributors (listed in alphabetical order): Gabriel Asfour, Joanna Berzowska, Katharina Bredies, Otto von Busch, Elise Co, Paola Guimerans, Aimee Kestenberg, Valerie Lamontagne, Lia, Jay Padia, Hannah Perner-Wilson, Cait Reas, Casey Reas, Bradley Rothenberg, Jeremy Rotsztain, Mika Satomi, Laura Siegel, Becky Stern, Lesia Trubat, Tania Ursomarzo, Rainbow Winters, and Adafruit Industries, Bare Conductive, Diffus Design Studio, Forster Rohner, Intel + Opening Ceremony, International Fashion Machines, Invent-Abling, Nervous System, Processing.org, and the Unseen. Their work is beautiful, evocative, and it is an honor to include their designs. We want to thank Kyle Li for his assistance with 3D printing, and for his enthusiasm and dedication to everything he commits to. Special thanks to Stephanie McNiel, our photographer and research assistant, whose attention to detail and incredible eye for the perfect camera angle have made the original photography in this book stand out.

We also thank the reviewers who helped shape the project from proposal to book: Margarita Benitez, Kent State University; Clare Culliney, Manchester Metropolitan University; Elaine Evans, University of Leeds; Melissa Halvorson, Marist College; Helen Koo, University of California, Davis; Huiju Park, Cornell University; Danmei Sun, Heriot-Watt University; Theresa Winge, Michigan State University; and Wendy Weiss, University of Nebraska.

Finally, we give thanks to our families: to Luis Cabrer, to Jonah Brucker-Cohen, Adrian Brucker-Cohen, and Avery Brucker-Cohen. None of this would have been possible without their endless patience and support. We also want to thank our parents, Margaritka and Yordan Genovi, and Linda and Yoshioki Moriwaki, and siblings Tomo Moriwaki, Tiffany Moriwaki Tesoro, Cristina Stephany, and Brian Moriwaki.

The Integration of Technology into Fashion

"It is difficult to give an exact definition of fashion because the word has had different connotations throughout history; the meaning and significance of the word have changed to suit the social customs and clothing habits of people in different social structures."

—Yuniya Kawamura

After reading this chapter you will be familiar with the following concepts:

- The longstanding role of material innovation in fashion
- The impact of technology on the design, manufacturing, and distribution process
- Wearables: the relationship of technology and the body
- Wearable computers, their history, and the stakeholders
- Computational and electronic fashion development
- The DIY (Do It Yourself) and Maker movements and their relationship to fashion

Technology has always impacted the practice of fashion design. From materials to new manufacturing processes, advancements in technological possibility have driven forward our understanding of the aesthetics, style, and functionality of clothing. Fashion as a discipline has always engaged with technology. With each new development there has been disruption, radical change, and innovation—in materials, manufacturing, and distribution.

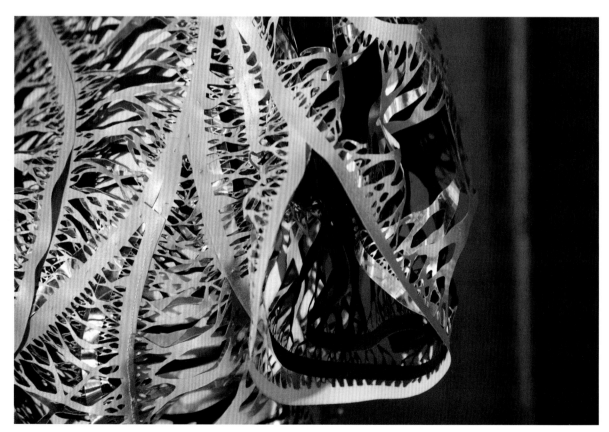

Figure 1.1 Detail of a laser cut leather dress by threeASFOUR.

We are at an exciting point in time where emerging technologies, computational and electronic design, and digital fabrication are creating new opportunities for designers. By understanding some of the ways in which fashion has already integrated technology into practice, we can gain insight into how specifically computation, electronics, and digital fabrication are both similar and divergent from previous developments.

Materials

The innovation of new materials has always had an influence on fashion. Historically such changes in materials have predicated new aesthetic interactions. From cotton, silk, and wool to bio-engineered leather, materials have affordances, which suggest aesthetic possibility for the designer. As available materials have changed, designers have adapted accordingly.

The advent of the industrial revolution saw the mechanization of yarn and eventually all textile production, starting with the Spinning Jenny, invented by James Hargreaves in 1764 (Figure 1.2). The device allowed one worker (then called a spinster) to produce up to eight spools of yarn simultaneously, as opposed to the single spindle on conventional spinning wheels.

This increase in output corresponded to the subsequent invention of automated textile looms,

Figure 1.2 Woman using a Spinning Jenny, c 1880. On James Hargreaves original machine, a single wheel controlled eight spindles. Later versions had upwards of eighty spindles.

Figure 1.3 The Swedish band ABBA often performed in spandex costumes during the late 1970s.

which could produce fabric in higher quantities than traditional hand weaving, culminating in the Jacquard loom, invented by Joseph Marie Jacquard in 1801, which could produce complex patterns such as brocade, damask, and matelassé. The Jacquard loom was controlled by punch cards, an inspiration for early pioneers in computing, as a means to "program," or provide, instruction sets to early computers.

Industrial age technologies fueled demand for mass-produced fabric, greatly increasing availability and choice of clothing for the lower and middle class while offering greater novelty to the elite. This material change in textile production, allowing greater availability and affordability, meant that fashions could diversify and new markets could develop, creating new opportunities for designers.

In 1935, Wallace Carothers created Nylon, the first synthetic fiber at DuPont's research facility at the DuPont Experimental Station. This invention lead to women's stockings, better known as "nylons," in the 1940s. In subsequent years several advances in synthetic fibers have been introduced, many with their own effect on ordinary habits of clothing and dress.[1] Polyester (introduced as Dacron in 1951) is the second most used fiber after cotton[2] and remains fixed in popular memory with 70s disco and polyester suits despite its widespread usage in contemporary clothing.

1. Susannah Handley, *Nylon: The Story of a Fashion Revolution: A Celebration of Design from Art Silk to Nylon and Thinking Fibres* (Baltimore: Johns Hopkins University Press, 1999).
2. A. Richard Horrocks and Subhash Anand, *Handbook of Technical Textiles* (Boca Raton: CRC/Woodhead, 2000).

Fabrics such as Lycra (also called Spandex (Figure 1.3) or Elastane), invented in 1959, revitalized the hosiery and lingerie market, preventing the sagging and bunching caused by nylon and creating moldable garments that moved with the body rather than restraining it. Initially found in active wear and associated in the 1980s with the popularity of aerobics, Lycra has now made its way into regular clothing and fashion styles, usually blended with natural fibers, an integral part of many everyday garments.[3]

Gore-Tex is another example of a synthetic fiber whose properties have influenced fashion. Invented in 1969, Gore-Tex is a waterproof, breathable fabric membrane often associated with outdoor and performance clothing, but also used extensively in industrial and medical applications. As a technical textile, Gore-Tex's ability to repel water while still allowing for the dissipation of sweat and heat from the wearer revolutionized outdoor and performance wear.

The process of creating technical textiles stems from needs found not only in fashion but also industrial applications in manufacturing, aerospace, and health and safety. Many advances originally intended as solutions for industrial applications have found unexpected applicability in fashion. Specifically, many conductive fabrics were originally used for electrical shielding before being used to create soft circuits that could be embedded into clothing.

In the present day, companies are experimenting with materials engineered with anti-microbial, stain-resistant, and thermal-regulating properties. These technical achievements have been a part of fashion's traditional way of engaging with materials, and many of these advances have already been integrated into designers' product lines. While computational technology and materials have yet to be used in the market on a large scale, the history of material invention suggests that this is a likely possibility as other technologies have proliferated in the past.

Process

New methods of design have created new forms of fashion. From sketching to pattern making and textile development, the use of multimedia software and various technologies has impacted the process of creating clothing. As the new tools change the process of designing fashion, the traditional production methods have to change as well. Designers are battling the need for low cost production and the export of labor outside of the US, while customers are increasingly demanding customization of goods and a faster production cycle. Thus the process is changing on a large scale at every level of design, production, and manufacturing through optimization with software and varied hardware technologies. Each stage becomes increasingly dependent on technology in order to meet the shorter deadlines and faster runway-to-store cycle.

Past innovations in process have had transformational effects on fashion design. The invention of the sewing machine, for example, allowed fashion to grow into a mass manufactured industry. Though there is debate about who deserves credit for inventing the sewing machine,[4] the result was a marked decrease in time required to sew clothes, which were up until that point done entirely by hand, and an increase in the availability and affordability of clothing for all social classes. Industrial age advances in technology built a mass market, diversifying the design process from custom tailoring available only for elites to ready-to-wear.

---

3. Kaori O'Connor, *Lycra: How a Fiber Shaped America* (New York: Routledge, 2011).

4. Grace Rogers Cooper, *The Invention of the Sewing Machine* (Washington, D.C.: Smithsonian Institution, 1968).

Another example of technological disruption from the industrial era is the invention of the modern sewing pattern, which offered fashionable styles in various sizes for home use. This was a serious technological innovation for its time, allowing the latest fashion to be reproduced and disseminated.[5]

In contemporary times, digital technology has already made an imprint on the design process. General graphics software such as Adobe Photoshop and Illustrator, used by designers in a wide range of fields, has been in use within fashion for quite some time. Software applications have become well-integrated tools for creative expression and rapid prototyping. Digital printing technology is already in wide use in the fashion industry, circumventing the limitations of large minimum yardage quantities and image engraving for textile printing, which is particularly attractive for either prototyping or the high-end market. 2D and 3D CAD apparel software have been in use since the 1990s with improvements to the technology that allow for the simulation of fit and drape in software. 3D body scanning has lead to more accurate modeling of the human body, allowing in some cases customized fitting at a mass scale. Developments in digital fabrication are a continuation of this trend, but also bring new possibilities to generate form algorithmically through code.

Distribution

The distribution and consumption of fashion has changed significantly with developments of technology. Social media sites and fashion blogging, online retailers, and the proliferation of fashion imagery have created an environment of perpetual consumption—if not of goods then certainly of the images that make up the world of fashion. Direct distribution to customers and on-demand customization has opened many small markets and created providers whose tastes would not have been accommodated in previous eras. As the ways in which fashion is consumed multiply, so do the channels through which designers can spread their vision and embrace the ever-changing environment.

In the past, one of the primary ways of distributing fashion was through replication achieved at home using the sewing machine and printed sewing patterns, mentioned in the previous section. This was eventually replaced by mass manufacturing through department and clothing chain stores. Computational technologies have the potential to use the mass communication channels of a networked society to enable a more nuanced combination of global distribution and mass customization. Designers and fashion companies working with digital fabrication are exploring this in new and exciting ways.

Technology on the Body

Technology and the body have always been intimately tied together. The amplification and extension of human faculties through body-mounted or worn apparatus goes back to the start of human history. Clothing, whether for physical protection and comfort or for personal adornment and projection of status, serves the vital function of mediating and modulating our presentation of self. More recently the integration of technology and the body have centered on computing and electronics.

Described loosely as **wearables**, this area has evolved to encompass technology devices (e.g., gadgets such as smart watches and head-mounted displays), computational and electronic fashion (such as the work by CuteCircuit[6]), as well as materials

---

5. Joy Spanabel Emery, *A History of the Paper Pattern Industry: The Home Dressmaking Fashion Revolution* (London: Bloomsbury Publishing, 2014).

6. CuteCircuit, accessed June 15, 2015, www.cutecircuit.com

science explorations of textile technology (Forster Rohner Textile Innovations (Figure 1.4)). That the term is so broad speaks to the rapidly evolving pace of innovation and the negotiation of disciplinary boundaries still taking place.

Within the realm of fashion and technology, wearables have always had a complicated relationship to fashion. One of the main criticisms of wearable computers is the clunky form-factor and cyborg appearance of early adopters. The unfashionable appearance of wearable computing **borgs** was seen as a primary hurdle to widespread public acceptance, as well as symbolic of how computational technology was out of touch with the emotive and expressive components of the human experience.

And yet, with other counteracting technological developments, mostly from multidisciplinary practitioners who blended the competencies of the technologist with design, new forms of wearables began to appear that challenged the conventionally accepted appearance and function of wearable computing. As a result this space is highly blended, with examples coming from diverse sources at the intersection of fashion, art, technology, and engineering.

The next three sections form a progression—historically, technologically, and sociologically—to the current intersection between fashion and technology. Without the success and early adoption of first attempts at creating functional wearable

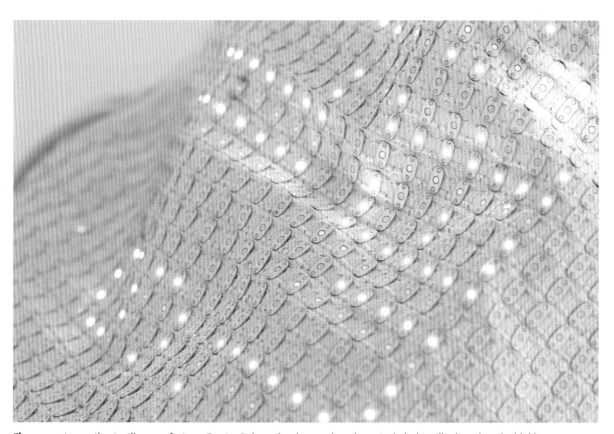

Figure 1.4 Innovative textile manufacturer Forster Rohner develops and produces technical textiles based on the highly sophisticated capabilities of modern embroidery techniques interconnected with electronic components on fabrics. Forster Rohner creates a unique blend of lights and textile with its proprietary e-broidery® technology and supports designers in all stages.

computing, we would not have the current state of technological and conceptual innovation.

Computational and electronic fashion challenged the affordances of what we think of as the traditional computer, providing an alternative that is aesthetically more aligned with fashion. The **DIY** and **Maker community** helped popularize hobbyist level tools and technologies to support laser cutting, 3D printing, and the use of microcontrollers in clothing, making these viable tools for many more people to use in prototyping and design.

Wearable Computers

Within the wearable computing community, eyeglasses and the wristwatch are often cited as accessories that have set the foundation for wearables as described by Bradley Rhodes on his website "A Brief History of Wearable Computing."[7]

7. Steve Mann, "A Brief History of Wearable Computing," accessed February 22, 2015, http://www.media.mit.edu/wearables/lizzy/timeline.html

While this does not fully capture the range of human accouterments that could be seen as body integrated technology, eyeglasses and the wristwatch represent the way in which bodies have always amplified and extended their senses through worn technology.

Seminal researchers in the field point to cognitive and sensory enhancement among the main benefits of the technology. Early versions of wearable computers were often considered clunky and obtrusive by the general public. The head-mounted display, chording keyboard, and fanny pack (containing the computer) heralded the advent of early adopters who called themselves "borgs." These early adopters were mostly computer science researchers for whom the benefits of having a mobile computer at one's fingertips outweighed the drawbacks of its highly visible and at the time socially unacceptable, extreme appearance.

Over the years the visual footprint of wearable computing has become much smaller (Figure 1.5). As with computer technology in general, smaller, faster processors and improvements in memory,

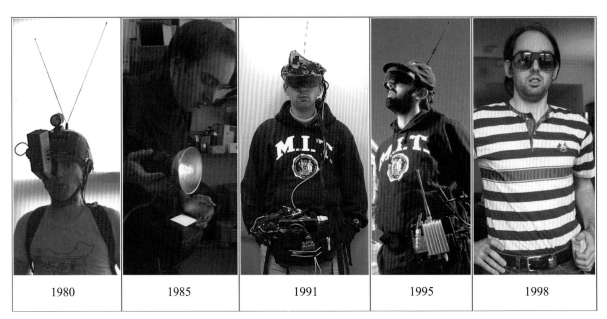

1980   1985   1991   1995   1998

Figure 1.5 Decrease in size of Steve Mann's wearable computing setup from 1980 to 1998. Technology had become less obtrusive but not necessarily more fashionable.

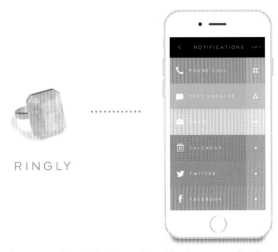

Figure 1.6 The Ringly electronic notification system is a line of connected rings that lets you put your phone away and your mind at ease. Ringly connects to your phone and sends you customized notifications through vibration and light.

Figure 1.7 The smart bracelet MICA was created by Opening Ceremony and engineered with Intel® technology, allowing you to easily view messages, your calendar, and alerts from your curated VIP contacts. Described by the company as "a feminine fashion accessory with communications capabilities," the bracelet was designed with 18K gold coating and a curved sapphire glass touchscreen display and was officially introduced on September 7 at the Opening Ceremony Spring/Summer 2015 fashion show in New York City.

battery consumption, and cost have brought the size of wearable computers more in line with conventional accessories. Rather than attempting to sell full service wearable computers (as defined by Steve Mann in his keynote address at the 1998 International Conference on Wearable Computing[8]), many companies are diffusing computational functionality into various enhanced accessories that take on various tasks an integrated wearable computer would offer while embedded within a bracelet (MICA by Intel and Opening Ceremony), rings (Ringly), or eyewear (Google Glass). For the designer interested in working in this area, development in this space is still dominated by technology companies, with a technology product development process, though this is changing as more fashion companies engage though partnerships or their own endeavors.

Computational and Electronic Fashion

In the late 1990s research and product surrounding wearable technology bifurcated, establishing a trajectory of work that placed the aesthetics of the artifact and experience at the forefront. An article published in the IBM Systems journal outlining the implementation of "e-broidery" by Post, Orth (Figure 1.8), Russo, and Gershenfeld[9] presented an early vision for computers that veered radically from the hard-angled boxes and stiff wires associated with electronic devices. The depicted work, which featured electronic components attached directly onto fabric substrates and examples of embroidered sensors sewn with conductive thread, was supple, soft, and revolutionary in terms of the design possibilities for the integration of computing into fabric and clothing.

8. Steve Mann, "Definition of 'Wearable Computer,'" (Keynote speech, 1998 International Conference on Wearable Computing) accessed June 15, 2015, http://wearcomp.org/wearcompdef.html
9. E. R. Post, M. Orth, P. R. Russo, and N. Gershenfeld, "E-broidery: Design and Fabrication of Textile-based Computing," (*IBM Systems Journal* 39, no. 3–4, 2000: 840–60) accessed August 2, 2015, http://gtubicomp.pbworks.com/f/post-isj393-part3.pdf

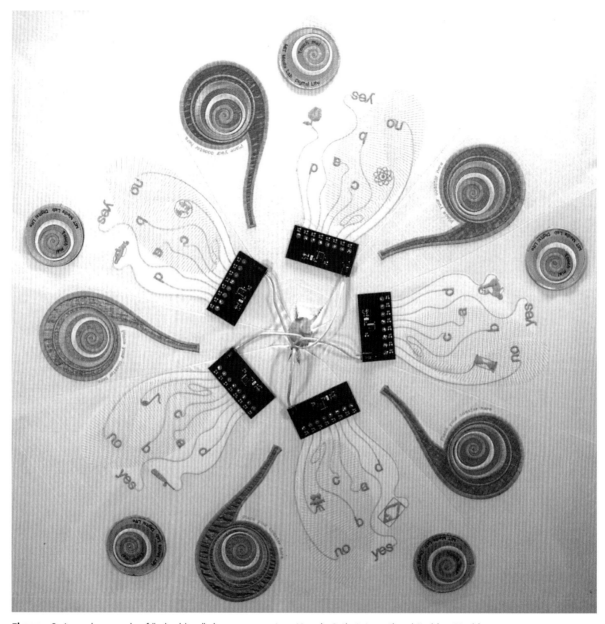

Figure 1.8 An early example of "e-broidery" circa 1997, courtesy Maggie Orth, International Fashion Machines.

At the same time the language and patterns of early interaction design began imprinting upon designers working in this space, most notably Hussein Chalayan and Walter Van Beirendonck. The architectural and kinetic quality of Chalayan's work, embodied by the Airplane Dress in 2000, with mechanical and electronic components integrated into the garments, was often cited as visionary from practitioners in a wide range of disciplines. As the creative director of the i-Wear project at the now defunct Starlab, Beirendonck focused on the functional as well as aesthetic

Design work done in this space is highly dependent on collaborative, multi-disciplinary, and creative production that either defies or straddles innovative boundaries. Whether in research labs such as the Intelligent Fibres group at Philips Design,[10] which developed conductive fabrics and wearable electronics for collaborations with fashion companies, or academic institutions such as the MIT Media Lab, or the work of independent creatives and designers, computational and electronic fashion points toward an operational shift in the kinds of behaviors and function traditionally associated with clothing and the act of "getting dressed." Computational technology is able to utilize the metaphor of "second skin," both figuratively and literally.

DIY and the Maker Movement

In January of 2005 the first *Make* magazine was published. Since then the **Maker movement** has become highly visible in mainstream cultural consciousness. Greater availability and affordability of fabrication technology has lead to a return to smaller scale, integrated production of materials and goods. Additionally, the lowered barrier of entry created by hobbyist level electronics and programming platforms has made building computationally augmented electronic objects more attainable.

Makers exemplify the hybrid craftsmanship of merging both traditional tools and approaches with emerging technology, often with an idiosyncratic aesthetic sensibility. Makers were often known, at least initially, for creating unusual electronic and digitally fabricated objects and devices for their own enjoyment. As time went on, exuberant displays

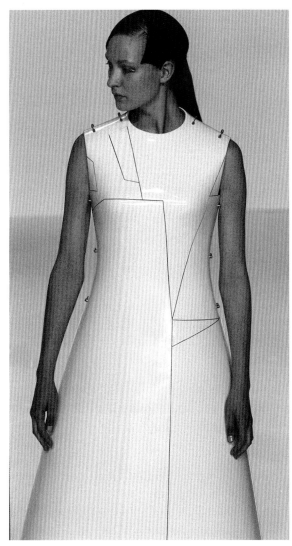

Figure 1.9 Hussein Chalayan's Airplane Dress inspired a large number of techies in the early 2000s to follow through on the dream of realizing computationally enhanced garments.

possibilities of new technology. Experts from a wide range of disciplines began to congregate in this space, many coming from non-traditional design and fashion backgrounds, resulting in a high variability of finishing in the final outcomes, at times drawing from the language of interaction design and software development with the release of beta versions and prototypes.

10. Stephano Marzano, Josephine Green, C. Van Heerden, and J. Mama, *New Nomads: An Exploration of Wearable Electronics by Philips* (Rotterdam: 010, 2000).

Figure 1.10 Electronic fashion from the Maker community, courtesy of Becky Stern and Adafruit Industries.

of novel making gave way to entrepreneurship in the form of small businesses, often targeted to a niche audience that could not be served by more expensive industrial processes. This led to an eventual overhaul of mass-market models as Maker-run businesses emerged to compete at a large scale.

Makers in Action: Popularizing Tools

The Maker movement has had a large impact on the relationship of technology to fashion, promoting wide adoption of the techniques and tools used to create computational fashion. A key tenet of making is the opening up of one's process, to share information in an "open source" or "share-alike" manner. Until Makers began publishing tutorials and supply lists online, detailing how to work with the popular electronics development platform Arduino, or how to sew with conductive thread, much of the specific practical knowledge for how to create interactive garments was shrouded in an aura of mystery behind the work of a handful of creators or hidden as intellectual property by large companies. Makers deconstructed these tools, identified how to source parts, and generally provided a template for building functional interactive garments by disseminating their knowledge. To this day the Maker community continues to evangelize in this area, most notably Leah Buchley, creator of the popular LilyPad Arduino, and Becky Stern (Figure 1.10), director of wearable electronics at the Maker-run company Adafruit.

One of the key benefits of cultural production in this area is the increased accessibility and visibility of tools and techniques for the designer. Though the professional fashion community may not engage heavily in DIY and Maker culture, and in some cases may even reject the comparison of home-crafted fashion goods to couture fashion, there is a rich tradition of skill-sharing and documentation that can benefit the designer in terms of understanding how electronic fashion and computational garments are built and prototyped.

The tremendous visibility of the Maker community and its enthusiasm for electronically integrated clothes has driven the popularity of off-the-shelf tools and kits that can be used by designers. The reduction of cost and the greater availability of digital fabrication technology and electronic prototyping tools are in part credited to the Maker community. For example, as demand for personalization and ownership of fabrication technology has boomed throughout the past few years, the cost of small portable 3D printers has fallen tremendously and it is now possible for individuals to equip their design studios and even their homes.

It has become evident that as a result of the efforts of the Maker community to openly share its tinkering and design process development, as well as the technology and techniques used for each project, fashion designers can benefit by learning these skills not traditionally taught in a fashion program or design studio. As technology is evolving to be embedded in clothing and wearables become synonymous with fashion accessories, designers increasingly find themselves in an unfamiliar territory and need to learn new skills associated with this new design thinking and process.

Design Aesthetics

As high fashion and the tech community are starting to converge, fashion has a great deal to contribute in terms of expertise and knowledge. The early adopters of wearable computing tended to exhibit a limited sense of aesthetics and inability to overcome the general population's lack of tolerance for overt displays of technology on the body. While interaction designers working with computational fashion introduced new materials and a reconsideration of the computer as a hard-edged device into something supple and tactile, many of these early efforts remained separate from the professional fashion industry.

The Maker community helped to popularize the tools and techniques of building interactive garments and accessories, which has increased interest and adoption of techniques and tools. Along the way, there have been visionary fashion designers who have seen the possibility afforded by emerging technologies, but much that has been done has remained conceptual and exploratory in nature. Increasingly, however, this is changing, with more interest and desire on the part of the designer to investigate possibilities created by the confluence of technology, interaction, and accessibility.

Although designers from the Maker and DIY communities have built garments and accessories, what is considered "on trend" is still managed with a strong degree of gatekeeping by those who work in the fashion industry. Fashion can be seen not only as set of actions (e.g., sketching, draping, sewing clothes) but also as a sociological construct.[11] The industry's definition of fashion does not include all people who merely make and create garments and accessories.

Working in multi-disciplinary teams, we are seeing more work utilizing digital fabrication tools such as 3D printing and laser cutting. Designers such as Iris Van Herpen (Figure 1.11) exemplify the possibilities of digital fabrication when turned over in service of the designer's vision. Computational

---

11. Yuniya Kawamura, *Fashion-ology: An Introduction to Fashion Studies* (Oxford: Berg, 2005).

Figure 1.11 Iris Van Herpen designs.

garments developed by sports companies such as Nike for fitness are leading the way in terms of engaging the public with the concept of wearing computers on the body.

It is a matter of time before this is accepted as pure fashion much in the same way that many functional textiles and fashion styles have become part of a luxury lexicon. Accessories, which incorporate technology devices, are on the cusp of readiness for the public. There is still a long way to go in the development of fashionable and functional implementations, which is why radical experimentation in this area and a willingness to take risks is still a part of working in this field. The competencies of fashion design are essential to making this vision a reality.

CASE STUDY
VEGA EDGE

Vega was founded by Angella Mackey in 2011. Its first product launched was the Vega One illuminated coat. Angella is a Canadian designer with a diverse background in new media art, electronics, fashion, and product design who also works as a designer and consultant for various types of functional clothes, including those for commercial space travel and medical care.

The Vega product offering has expanded to include a range of outerwear with embedded electronics for fashionable urban cyclists. In its essence, Vega makes lights you can wear that complement the clothes you already have. Their functionality ranges from providing safety while cycling, running, or walking to blending in with your existing wardrobe as a fashion item with simple technology.

In 2013 Vega collaborated with Kate Hartman's Social Body Lab to create the Vega Edge (Figure 1.12)—a wearable light accessory, which had a successful Kickstarter campaign in March 2014. The concept is that wearing light can be part of everyday fashion if there's a clear purpose and simple execution without clutter and special

Figure 1.12 The Vega Edge clips on to the edge of a jacket or any other garment and illuminates to make the wearer more visible.

effects. The wearer can clip on a light that can be integrated into a garment to make the user more visible to oncoming traffic.

The design team believes that if technology is to be wearable, they should pay attention to fashion trends, materials, and design processes meant for making clothes. For this product the Vega team looked into old technologies and new execution techniques to balance the latest fashions with this added function.

Figure 1.13 Each Vega piece is laser cut from high quality durable leather.

Figure 1.14 The resulting pattern pieces are ready for the embedded technology.

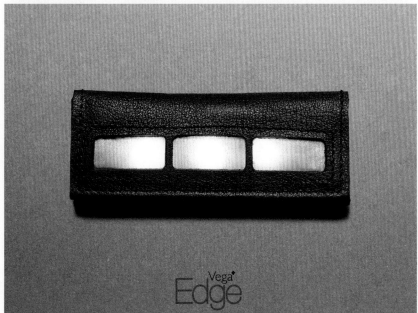

Figure 1.15 The finished Vega Edge illuminated product features various colors and shapes.

INTERVIEW
KATE HARTMAN (COLLABORATOR, VEGA EDGE)

Kate Hartman is an artist, technologist and an educator. Her areas of expertise are physical computing, wearable technology, and conceptual art. She is currently a professor of wearable and mobile technology at the Ontario College of Art and Design in the Digital Futures program.

Kate uses simple, open-source technology to build objects and DIY kits, such as her Inflatable Heart or Glacier Embracing Suit. She is the co-creator of Botanicals, a system for letting plants tweet and call their owners when they need water or more sunlight. Hartman is also the un-director of ITP Camp, a summer program for grownups at ITP/NYU in New York City. Her work is included in the permanent collection of the Museum of Modern Art in New York.

Figure 1.16 The Vega Edge team in their Toronto studio.

AG, KM: You are currently Capitalize: Associate Professor of Wearable Technology and capitalize Director of the Social Body Lab at OCAD University. What is your background and how did you get involved in wearable technology and fashion?

KH: I originally came from a background in photography, film, and video but later found myself drawn to more interactive and people-centric work. Growing up I had learned a bit about sewing and textiles from my mother who is a fiber artist. When I learned about electronics and programming in graduate school at ITP/NYU I was able to put it all together to create interactive projects that live on the body. Now, many years later, wearable technology has become my area of expertise.

Through my teaching at OCAD University I work with artists and designers and help them learn the skills that they need to create wearable technology and physical computing projects. The Social Body Lab is a research group that I had the fantastic opportunity to be able to create from scratch. We focus on the exploration and development of body-centric technologies in the social context. Our work engages with wearable technology from both a creative and a critical perspective.

AG, KM: Many of your projects are worn on the body and explore both personal expression as well as communication. What is the role of fashion in your work?

KH: Fashion is very much about personal expression and also how we connect with other people. I feel like the way that I work with wearable technology is just an extension of that. Both our body language and our clothing communicate so much. Adding some new tools to the mix creates an opportunity to expand the possibilities for that communication.

AG, KM: Tell us about the Vega Edge. How did you and your collaborators come up with the concept? How did you get involved in the project?

KH: I met Angella Mackey at an art opening when I first moved to Toronto in 2009. At the time she was the only person that I knew there who was working with fashion and technology. Together we founded the Toronto Wearables Meetup and from there a whole community started to emerge.

A few years later we received funding that enabled us to collaborate on a project. Angella had been working on a line of stylish jackets for cyclists that featured integrated lighting. She had many fans of her work but the price point of these jackets was a bit too high for most. We decided to work together on a more modular, affordable approach to wearable light that fit the look of her existing line. What we created held all of the concepts behind her jackets without having the added complication and costs associated with garment production.

AG, KM: What are the backgrounds of your collaborators? How did they get involved in wearable technology and fashion?

KH: Angella has a background in new media art and interaction design. This lead her to discover artists like Joanna Berzowska and Valérie Lamontagne, who she saw doing really interesting things with electronics and textiles in the early 2000s. She was then inspired to pursue fashion design so that she could integrate electronics into clothes from the ground up and have full control over the garment designs. She now works with this almost exclusively.

My team at the Social Body Lab is comprised of current students and recent graduates from OCAD University who come from a variety of disciplines. The lead research assistants on the Vega Edge project were Hillary Predko and Jackson McConnell.

Hillary is an interdisciplinary designer who got her start in Toronto's indie fashion scene. She learned wearable electronics techniques while at the OCAD and became fascinated with incorporating circuits and mechanics into textiles. She also was our laser-cutting expert throughout the Vega Edge production process.

Jackson had a background in industrial design and came to OCAD to pursue an interest in creative computing. His research focuses on technologies that live on or around the body—particularly how they mediate social interactions between people.

Beyond that we had many other fabulous collaborators, including David McCallum, Johannes Omberg, and Erin Lewis, who assisted with the technical development of the Vega Edge.

AG, KM: What was your experience working with fashion designers on a wearable technology design project?

KH: Fantastic! I do not have any training in fashion so the gain for me is significant. I find it important to work with fashion designers who either have some knowledge of or a strong interest in technology and its possibilities. When technology is to be included in fashion design, the process of integration needs to start at the very beginning. This means that fashion designers need to be willing to open up, share, and adapt their processes.

AG, KM: What was the philosophy behind the Vega Edge? How does the design process you engaged in reflect that?

KH: There were several key factors that we considered when creating the Vega Edge. The first was that it needed to be modular. As people we wear lots of different clothing. The problem with technology that is designed into a specific garment is that there is a limit to how much use you can get out of it. We wanted to make wearable light that could be added to any garment—a light module that could be easily attached and detached so that you could put it anywhere you like.

Another factor that was considered was what Angella calls the "Off Value." This calls the question of whether the accessory looks good even when the technology isn't in action. Particularly when working with light it can be easy to get carried away with how interesting the object looks when it's shining and blinking. However, we put just as much (maybe even more) thought into how it might blend with a person's wardrobe for the majority of the day when it wasn't being used as a light. Amongst other things, this led to the main material choice of leather—a high-quality, durable, and very common material that already has a culture and history of pairing with clothes. This helped bridge the gap between strange new object and something people could see themselves wearing.

AG, KM: What was the design process that went into the Vega Edge? The project seemed to pull together various emerging technologies as well as technological processes. Can you speak about those technologies and processes both independently and as a whole in terms of how they worked together to create the finished product?

KH: Our initial process was very exploratory. We looked at fashion magazines and blogs for inspiration on the different forms it could take. We experimented with many materials, including leather, wood, and silicone. Fabrication techniques included mould making, laser cutting, and 3D printing. We also spent a lot of time exploring different methods for attaching the wearable to different clothing and accessories, and testing out our rapid prototypes by taking them out for a spin at night.

These experiments helped us make design decisions and move towards a solid first prototype—a light enclosed in a laser-cut leather case that used strong rare earth magnets to clip to the edge of any piece of clothing. From there we were able to start iterating. The laser cutter proved to be our most valuable tool, allowing us to make many, many versions of the design quite quickly, better enabling our process of moving the design forward.

Finally, we had to consider the way it would be manufactured and the final cost. This meant stepping outside of our comfort zone of DIY and imagining how hundreds of copies could be produced outside of our studio setting and with careful consideration of our suppliers. This took a lot of research and ultimately had a huge impact on the final piece as well.

AG, KM: What do you see as the main challenges and opportunities facing young designers interested in this area of practice?

KH: Wearable technology sits at the intersection of a variety of disciplines—engineering, computer science, fashion design, industrial design, and many others. Some traditional education systems make it difficult to engage with such a broad diversity of subjects. However, there [are] a growing number of interdisciplinary programs that are starting to make this a bit easier.

AG, KM: What advice would you give to someone who is just starting out in this area?

KH: Keep learning and always collaborate! There are always more skills and knowledge to be gained. But you can't know everything. Working with people who have a deep knowledge in a subject area that is new to you can be both great for your project and an opportunity for you to further your learning!

INTERVIEW
JOANNA BERZOWSKA

Joanna Berzowska is Associate Professor of Design and Computation Arts at Concordia University and a member of the Hexagram Research Institute in Montreal. She is the founder and research director of XS Labs, where her team develops innovative methods and applications in electronic textiles and responsive garments.

Her art and design work has been shown in the Cooper-Hewitt Design Museum in New York City, the V&A in London, the Millennium Museum in Beijing, various SIGGRAPH Art Galleries, ISEA, the Art Directors Club in New York City, the Australian Museum in Sydney, NTT ICC in Tokyo, and Ars Electronica Center in Linz, among others.

She lectures internationally about the field of electronic textiles and related social, cultural, aesthetic, and political issues. She was recently selected for the Maclean's 2006 Honour Roll as one of "thirty-nine Canadians who make the world a better place to live in."

She received her Masters of Science from MIT for her work titled "Computational Expressionism" and subsequently worked with the Tangible Media Group of the MIT Media Lab and cofounded International Fashion Machines.

Figure 1.17 The Shoulder Dress is part of Joanna Berzowska's Karma Chameleon research/creation project, a collaboration between a designer and a scientist. This garment is constructed out of a new generation of composite fibers that harness power directly from the human body, stores that energy, and then uses it to change its own visual properties. This animated garment changes its visual characteristics, color, and shape in response to physical movement.

She holds a BA in Pure Math and a BFA in Design Arts.

AG, KM: Your background and education is in both math and design arts. What led you to the fields of electronic textiles and wearable computing?

JB: In high school, I was really good at math but also really loved visual arts, design, and theater. When I graduated from high school in 1989, I decided to pursue two different degrees, one in math and one in design, at two different universities, at the same time. It was definitely unusual, but in a way, I created my own education to become a designer with expertise in science, mathematics, and programming. This is exactly the way that we now teach in my Department of Design and Computation Arts at Concordia University.

At the time, however, there were no degrees in multimedia or interaction design, so I built my

Design Aesthetics

own hybrid education by studying both math and design. My interest in electronic textiles and wearable computing, therefore, emerged from my interest in science and all these new computer technologies and how they could be applied to design. I especially wanted to develop provocative work in the field of interaction design, which was obsessed with "hard" objects (the mouse, the keyboard) and limited interactions, by introducing "soft" interfaces in the form of textiles with sensors and the possibilities for creating digital interaction on the whole body through wearable computing.

AG, KM: What was the design philosophy at XS Labs, which you founded in 2002?

JB: A lot of the work happening with wearable computing, through the computer technology community and specifically HCI (Human Computer Interaction), focused on tangible interaction and involved the manipulation of physical objects with the human hand. I anticipated that electronic textiles would allow us to expand the realm of physical interaction into a truly wearable context and to explore the boundaries of what I call "beyond the wrist" interaction. I love the idea that we can have computer interfaces that work on our whole bodies, through interactive garments.

There are so many interesting new materials for designers to use, such as conductive fibers, active inks, photoelectrics, and shape-memory alloys. They promise to shape new design forms and new experiences that will redefine our relationship with color, texture, silhouette, materiality, and with digital technology in general. My design philosophy at XS Labs focuses on the use of these active, reactive, and interactive materials and technologies as fundamental design elements. I focus on the aesthetics of interaction to question some of the fundamental assumptions we make about the technologies and the materials that we deploy in our designs. I love to play with unexpected material changes, such as a textile design that disappears because it was printed with thermochromic ink, a dress that moves up and down on your body because of the shape-memory alloy fibers . . .

AG, KM: Much of the work of XS Labs focuses on the expressive potential of textile and wearable technology. How does this contrast to the more functional orientation of current wearable technology?

JB: My projects at XS Labs try to take a different approach than the traditional task-based, utilitarian definitions of functionality in HCI. My definition of "function" simultaneously looks at the materiality and the magic of computing technologies; it incorporates the concepts of beauty and pleasure. I am particularly concerned with creating wearable computing that is surprising, compelling, strange, and beautiful, rather than wearable computing that reads your e-mail or tells you the weather . . . We have to stay true to the history of fashion and elevate fashion through electronic technologies, rather than replacing fashion with consumer electronics.

AG, KM: What is missing in the world of wearable and portable electronic devices?

JB: We need wearable computing that is irrational, poetic, musical, and theatrical. We need wearable computing that stimulates magical and literary experiences in our everyday life rather than just trying to increase our productivity or our efficiency. Many of my electronic textile innovations are informed by the technical and the cultural history of how textiles have been made for generations—weaving, stitching, embroidery, knitting, beading, or quilting—but use a range of materials with different electro-mechanical properties. I consider the soft, playful, and magical aspects of these materials, so as to better adapt to the contours of the human body and the complexities of human needs and desires. My approach often engages subtle elements of the absurd, the perverse, and the transgressive. I construct narratives that involve dark humor and romanticism as a way to drive design innovation. These integrative approaches allow me to construct composite textiles with complex functionality and sophisticated behaviors.

AG, KM: Should all our clothes and accessories be connected or do you think we need to disconnect a bit too?

JB: Ten years ago, when I used to travel almost every week for lectures or research, I especially loved the travel itself because it forced me to be disconnected. We did not have WiFi on airplanes and I did not have e-mail on my phone. . . . Those hours of silence, of solitude, were very luxurious and very productive. I was able to be alone with my thoughts, to develop new ideas and new projects. Now that connectivity has become more ubiquitous, it almost feels like a wearable that forcibly disconnects you would be a really provocative and a really uncomfortable thing. I like that! I think it is important for us to be placed in situations that are difficult, to force us out of our comfort zones, to stimulate learning and growth.

AG, KM: Please tell us about your latest project you are working on.

JB: I am the head of electronic textiles at OMsignal, a Montreal startup developing wearable technology products that focus on performance, wellness, and well-being. Responding to the growing need in our society to find balance in our lives, the first product is a shirt that tracks various bio-signatures through textile-based sensors and through the iPhone, and offers a variety of engaging biofeedbacks to help improve performance, well-being, increase self-knowledge, and reduce stress.

AG, KM: What do you think are the greatest achievements and technological challenges in the development of smart textiles?

JB: Manufacturing remains one of the greatest issues. . . . Innovative design concepts are difficult to manufacture at scale in an affordable manner. The problem is that there are no existing large-scale manufacturing capabilities for electronic textiles, as well as the difficult integration of textile and electronic technologies, since most large manufacturing operations focus either on textile or on electronics with little to no overlap and small profit margins. We still have to solve the considerable reliability and scalability issues in manufacturing.

AG, KM: What inspires your work first and foremost, form or function?

JB: I believe that form *is* function! Smart garments integrate new technologies that provide added value to people, creating intermeshed links between our physical and digital identities, as well as our social networks. Clothing is meant to work with our bodies, not just decorate them, helping to mediate our individual and social identities, increase self-awareness, and improve our quality of life.

AG, KM: Your work involves a lot of multidisciplinary collaboration. Could you elaborate on what that experience is like? What are the challenges and opportunities associated with working between disciplines?

JB: In the last five years, I have been working with scientists to develop a new generation of composite fibers that have computational functionality. The core technical innovation involves shifting this functionality entirely within the fiber itself. The goal of this project, entitled "Karma Chameleon" (Figure 1.17), is to develop a prototype for an all-fiber based textile that can harness, sense, and display energy. Conceptually, this constitutes a radical deviation from the dominant model of a textile substrate with integrated mechano-electronics to a fully integrated composite substrate, wherein the fibers themselves (a) harness human-generated energy, (b) store the energy directly inside the fibers, and (c) use that energy to control a fiber-based actuator (such as fiber illumination and color).

The design implications of such new fibers are twofold. First of all, when a material integrates computational behavior, how do we program such a material? We do so by determining the length, the shape, and the placement of the material in a composite system (in this case, the textile). Changing its shape or orientation will change its behavior, not only visually, but also computationally. The second, more profound, implication is that the language of aesthetics and design (parameters such as shape, color, or visual composition) becomes conflated with the language of programming. So again, form *is* function!

AG, KM: What advice would you give to aspiring wearable tech designers?

JB: I think it is really important to have a passion for other fields, to be inspired by science, technology, and literature. . . . You have to try to blur the lines between technology, poetry, and materiality so as to develop deeper mastery of interactive, changeable design.

For Review

1. Why was the invention of Nylon so important to the field of fashion technology?
2. How do you describe "wearables"? What is the relationship between wearables and fashion?
3. Name some of the early researchers and designers who implemented electronics into their work. What was groundbreaking about their work and how did it lead the way for further innovation?
4. What are some of the contributions of the Maker community to the fashion design community?
5. What is the role of fashion design aesthetics for wearables?

For Discussion

1. Name some of the latest innovations in wearables and discuss their design aesthetics versus functionality. What does a successful implementation of functionality look like?
2. What do you see missing from the functionalities?
3. How could you improve upon the current wearable designs offered?
4. When you search for DIY and Maker community projects, what do you find helpful?
5. If you were developing a fashion and tech garment and accessory what kind of people would you want to collaborate with?

Key Terms

DIY: Commonly used short term for "Do It Yourself." It defines the method of building, modifying, or repairing something by yourself, without the aid of experts or professionals.

Maker community: The group of people around the world who are becoming influenced to be DIY makers or are actively participating in the maker movement.

Maker movement: A contemporary culture or subculture representing a technology-based extension of DIY culture. Typical interests enjoyed by the maker culture include engineering-oriented pursuits, such as electronics, robotics, 3D printing, and the use of CNC tools, as well as more traditional activities such as metalworking, woodworking, and traditional arts and crafts.

Borgs: Early adopters, who were mostly computer science researchers, for whom the benefits of having a mobile computer at one's fingertips outweighed the drawbacks of its socially unacceptable appearance.

Wearables: Eyeglasses and the wristwatch are often cited as the first "wearables" in the history of wearable computing, as described by Bradley Rhodes on his website "A Brief History of Wearable Computing."

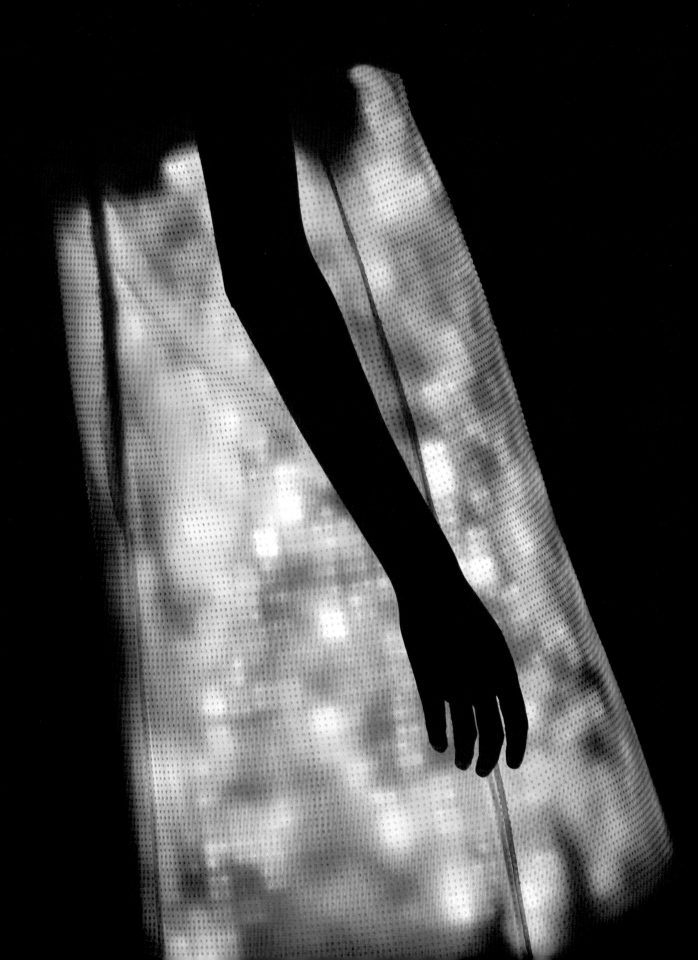

2

How to Design with Electronics

This chapter introduces the principles of electricity and essential concepts for basic circuits. It starts with fundamental definitions of basic components and builds a deeper understanding of how to build electronic circuits in the context of designing for the fashion and accessory industry. This knowledge is the basic building blocks for designers like you, who desire to integrate electronics into fashion. You will learn how to construct soft circuits and how to communicate your design ideas with the necessary terminology. The tutorials at the end of the chapter will build the skills you need to start your own designs with soft circuits. The Case Study of a flip switch by Katharina Bredies reveals a functional application that can be embedded in a garment or interactive project.

After reading this chapter you will:

- Know the basic principles of electricity

- Have the knowledge to build an electric circuit

- Be familiar with the basic components of a circuit

- Be able to build your own soft circuit

Learning the basic concept of electricity and circuits will inform your design process of integrating electronics into your garments and accessories, but keep in mind that your collection is driven by your main concept as a fashion designer. Thus, you evaluate how electronics will fit in to enhance your ideas. Learning the various kinds of soft circuits and components and how they can be integrated into your designs will be key in creating a successful and functional collection. The following questions should lead your work as a designer:

- How do I create a closed circuit in which electricity will flow uninterrupted?
- Would a series or a parallel circuit work better for my design?
- How do I power the circuit and how can the power source be integrated into the garment or accessory that I am designing?
- Am I allowing for functional storage where the customer can change the batteries or is the source self-powering?
- How do I turn on and off the circuit?
- How can I integrate the function of closing and opening the circuit into the garment?
- Which components will work best with the chosen materials in my collection?
- How will the switches in my circuit affect the overall design?

Electricity

First and foremost it is important to understand what electricity is and how it works. There are two types of electricity: **AC and DC current**. AC means that the current is alternating, and DC means that the current is direct. An **electric current** is a flow of electric charge, traveling from a point of high electrical potential, identified as *power* with the symbol for plus (+), to the lowest, which is usually identified as *ground* with the symbol minus (–). In electric circuits, this charge is usually carried by moving electrons in a wire, but in fashion we can use a variety of conductive materials, such as thread, paint, or tape, in order to achieve a desired look or simply hide the circuit itself.

When the current is alternating (AC current), the direction of the electricity flowing throughout the circuit is constantly interchanging direction. The rate of this reversal is measured in **Hertz (Hz)**, which is the unit of measure for frequency named for Heinrich Rudolf Hertz, who was the first to conclusively prove the existence of electromagnetic waves. One Hz simply means "one cycle per second." DC—or direct current—electricity flows in one direction between power and ground. In this scenario there is always a positive source of **voltage** and a negative or ground source of voltage, also called zero voltage.

Electricity has a voltage and a current rating. Voltage is the difference in electrical energy between two points and is rated in Volts (V). It is named after the Italian professor Alessandro Volta, who used zinc, copper, and cardboard to invent the first battery. Volta's battery produced a reliable, steady current of electricity. **Current** is the amount of electrical energy passing through a particular point and is rated in Amps (A) or Milliamps (mA). The unit of amp, which is short for *Ampere*, is named after André-Marie Ampère, a French mathematician and physicist, who played a big role in discovering electromagnetism and is considered to be the father of electrodynamics. An amp is one of the seven base units, according to the International System of Units. It is a measure of a constant current, or the nominal flow of charge per second through a simple circuit. To understand these definitions better, consider a brand new 9V battery. This battery would have a voltage of 9V and a current of around 500mA (500 milliamps).

Electricity can also be defined in terms of resistance and watts. In order to define what resistance is we should define first a **conductor**. A conductor is an object, material, or fabric that permits the flow of electric charges in one or more directions. The **electrical resistance** of an electrical conductor is the opposition, or resistance, to the flow of an electric current through that conductor. The international unit of electrical resistance is the **ohm (Ω)**, named after German physicist Georg Simon Ohm, who found that there is a direct relationship between the potential difference (voltage) applied across a conductor and the resultant electric current. Ohm's law states that Voltage (V) is equal to current (I) times resistance (R).

$$V = I \times R$$

Ohm's law equation can be rearranged and expressed in a couple more variations:

$$I = V \div R$$
$$R = V \div I$$

You can use Ohm's law to determine how much resistance is needed in a circuit with any number of components and to see if any LED, for example, is receiving more than the desirable amount of current.

In electric circuits it is also important not to exceed the wattage rating of a component or it can be damaged. **Watt** is the electrical unit of power and is named after James Watt, a Scottish inventor who made improvements to the steam engine during the late 1700s and helped jumpstart the Industrial Revolution. The wattage of any DC power supply can be calculated by multiplying the voltage and current of the power source.

Electric Circuit

A **circuit** is a complete and closed pathway or a never-ending loop through which an electric current can flow uninterrupted. A closed circuit allows the stream of electricity between power and ground, while an open circuit breaks the flow of electricity between power and ground. Anything that is part of this closed system and that allows electricity to flow between power and ground is considered to be part of the circuit. Each circuit has various components, which allow the electricity to flow and perform a predetermined function.

Each circuit should be tested before it is realized on the actual garment or accessory. There are various open source websites that act as online schematic circuit simulators. Websites such as Fritzing.org, Circuitlab.com, and dcaclab.com allow you to build, test, document and share your prototypes before you move onto the final garment.

Keep in mind you can create your own circuits to be as complicated, as whimsical, or as sophisticated as you like. For example, Paola Guimerans, who is a graduate of the MFA Design and Technology program at Parsons The New School for Design, has been crafting technology and exploring new expressions of interactive art and interdisciplinary educational

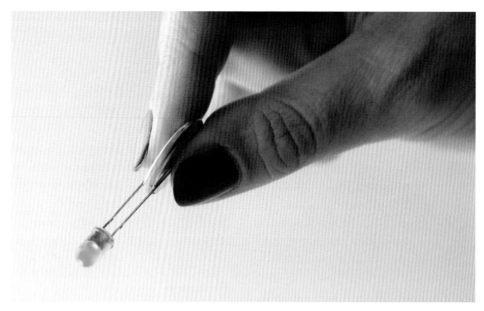

Figure 2.1 One of the simplest electric circuits: an LED powered by a battery.

approaches by combining visual arts, handcrafts, and electronics. Her embroidered circuit (Figure 2.3) utilizes traditional machine embroidery and is made from colorful top thread and a conductive bobbin thread underneath in order to complete the circuit. You could use the conductive thread as the main thread or use it only on the bottom, as Paola did in this circuit. This technique gives you the freedom to maintain a color palette that works better with your design.

Next we will review some of the basic and necessary components for a circuit.

Series vs. Parallel

There are two different ways in which you can wire a circuit. One is called series and the other parallel. When components are wired in series, they are positioned one after the other (Figure 2.4 right). That means that electricity has to pass through one component first, then the next one, and the next one, and so on. When components are wired in parallel, they are positioned side by side, so that electricity passes through all of them at the same

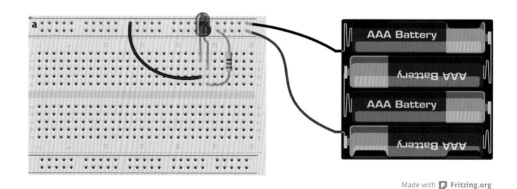

Figure 2.2 a One LED electric circuit shown as (a) schematic created on Fritzing.org

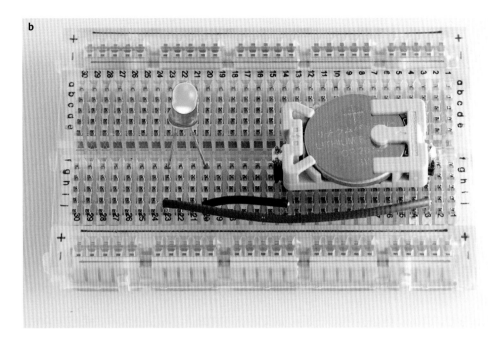

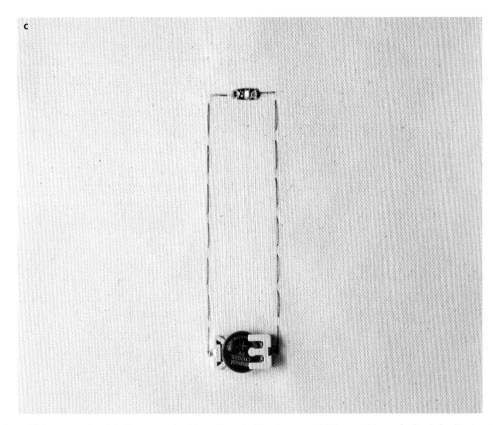

Figure 2.2 b, c (b) the same circuit built on an actual breadboard with wires, and (c) the resulting soft circuit finalized on fabric.

Electric Circuit

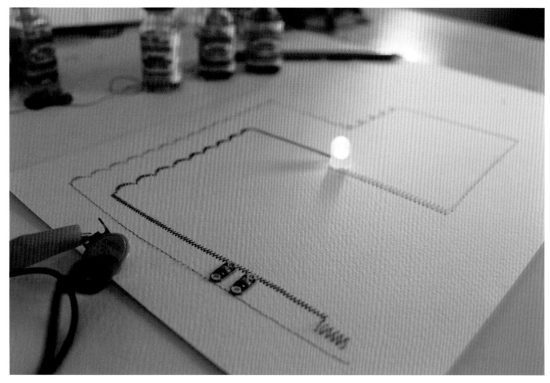

Figure 2.3 An embroidered soft circuit by Paola Guimerans with a combination of sewable and regular LEDs in parallel. Note that the bobbin thread (not seen on top) needs to be conductive in order to let electricity flow.

time (Figure 2.4 left). Keep in mind that power diminishes as it goes through the circuit in series. You should build a circuit in parallel if you want to have multiple LEDs.

Flexible Circuits and Electronics

Flexible circuits, or flex circuits (Figure 2.5), are one of the most important and fastest growing segments of electronics. A **flexible circuit** in its purest form is a vast array of conductors bonded to a thin dielectric film. The desired electronic components can be mounted on flexible plastic substrates such as polyamide, polymer thermoplastic or transparvent conductive polyester film. Flex circuits can also be screen printed as silver circuits on polyester, and flexible foil circuits or flexible flat cables (FFCs) can be created by laminating an ultra thin copper strip in between two layers of PET (Polyethylene terephthalate). This allows the board to adapt to any shape and be flexible during its use, and replaces any bulky wiring with an ultra thin connection. The size and weight is reduced to a minimum while we benefit from gaining flexibility and stronger signal quality. Mechanical connectors are eliminated, which reduces errors in wiring and connectivity.

Flex circuits are best used in wearable technology and accessories that have minimal surface area and must conform to the body shape or a size limitation. This technology has been used since the 1950s in electronics, the military, and the medical field, but it is rapidly advancing. Medical devices and cellphones have already benefited tremendously. Fashion clothing and accessories can also benefit from this technology because circuits can be added with minimal visibility and maximum flexibility. Flexible circuits can be designed and silk screened

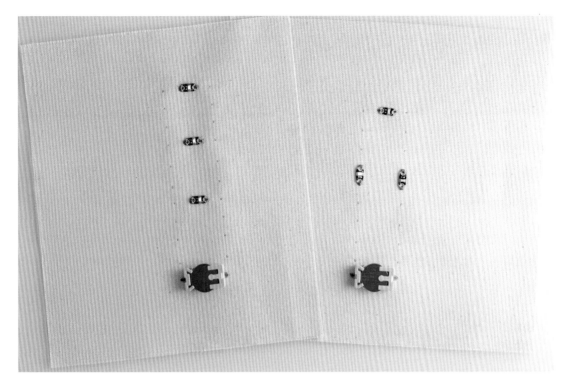

Figure 2.4 Photos of soft circuits with three LEDs: Left circuit shows three LEDs connected in parallel, and right circuit shows three LEDs connected in series.

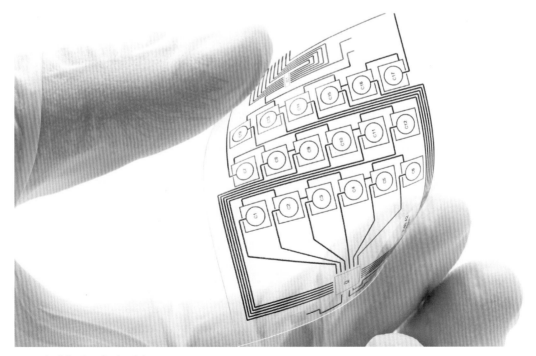

Figure 2.5 Flexible electric circuit layout.

Electric Circuit

with copper or silver inks on textiles to create a visually stimulating design.

Basic Components of a Circuit

In order to build a circuit, you need to connect a few components into a loop so that electricity can flow through it. The types of components you use will be determined by the function and look you would like to achieve. In a basic electronic circuit an engineer might use a wire, while a fashion designer might use a conductive thread or paint instead. A traditional button switch can be replaced with a snap or a metal zipper. Each component should be carefully selected based on its functionality and purpose. Some of these include batteries, LEDs, resistors, capacitors, diodes, transistors, potentiometers, switches, wire or conductive thread, conductive paint, and conductive tape.

Batteries

An electric battery is a container that stores power and can generate electricity to power the circuit. In its essence it is a device that converts chemical energy into electricity, which allows limited current

Figure 2.6 You can choose from a variety of power sources, depending on your particular needs. Shown in this photo are solar panels, lithium ion polymer batteries, thin flat coin cell battery, AAA alkaline, AA alkaline, and 9V alkaline batteries.

to flow out of the battery into the circuit. Each cell contains a positive terminal, or **cathode**, and a negative terminal, or **anode**, which is clearly marked on the outside of the container. There are two types of batteries: single use and rechargeable. However, even batteries of the same type can vary slightly in their voltage and capacity.

You can add batteries in series or in parallel. By placing batteries in series you are adding the voltage of each consecutive battery, but the current stays the same. For instance, an AA battery is 1.5V. If you put three in series, it would add up to 4.5V. If you were to add a fourth in series, it would then become 6V. By placing batteries in parallel the voltage remains the same, but the amount of current available doubles. This is done much less frequently than placing batteries in series, and is usually only necessary when the circuit requires more current than a single series of batteries can offer.

Resistors

A **resistor** is a component that adds resistance to the circuit and reduces the flow of electrical current (Figure 2.7). The traditional store-bought resistors have various markings that represent different values of resistance measured in ohms. Their values are read from left to right towards the gold band, and they can easily be interpreted by looking at a resistor color value chart. Resistors can be fixed or variable. As the names suggest, a **fixed resistor** has a predetermined value that cannot be changed, and a **variable resistor** can change its values depending on outside influence. It may have multiple touch points so that the resistance changes by moving the connection to different terminals, or there may be a slide along the resistance element that allows a larger or smaller part of the resistance to be used. When creating wearable tech or a garment, you may benefit

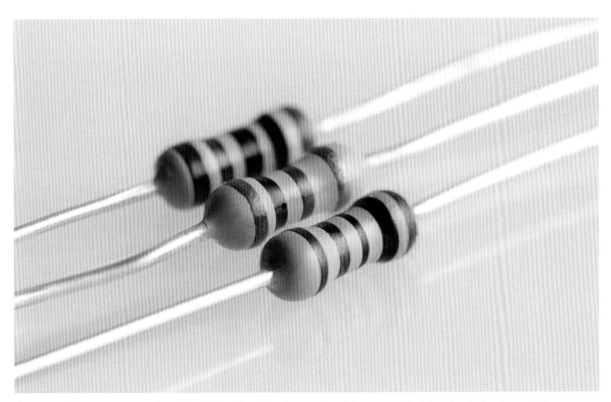

Figure 2.7 The arrangement and number of the colored bands on each resistor are used to determine their value and tolerance.

from making your own resistors from various fabrics. For example, a stretchy conductive fabric can create a functional variable resistor.

Photoresistors

A **photoresistor** or light-dependent resistor (shortened to LDR) or CdS cell (if it is made from Cadmium Sulfate), also called a photocell, is a variable resistor controlled by light (Figure 2.8). Its resistance decreases as it is exposed to increasing light intensity, and it exhibits photoconductivity. Photoresistors are small and very low cost and can be found in camera light meters, streetlights, clock radios, and solar lamps, but they can also be used in clothing or accessories as a switch for light- and dark-activated circuits. The resistance range and sensitivity of a photoresistor can differ greatly from one component to another, and there are many types of photoresistors used in various ways, but in general they tend to be inaccurate and should be used for basic detection of light and dark instead of exact values. They can also be coated with diverse materials that vary the resistance, depending on the use for each LDR, but keep in mind the use of photoresistors is severely restricted in Europe because of the Restriction of Hazardous Substances Directive ban on cadmium.

Potentiometers

A **potentiometer** is a variable resistor with two or three terminals and a sliding or otherwise movable contact element, called a wiper, which forms an adjustable voltage. The wiper slides across a resistive strip to deliver an increase or decrease in resistance. The level of resistance will determine output of current to the circuit. Traditional potentiometers look like a knob that you can turn and switch between terminals to vary the resistance (Figure 2.9 a, b).

Linear potentiometers are more commonly used in soft circuits. They have a wiper that slides along a linear element instead of rotating between terminals—or you can use your finger as the wiper (Figure 2.10 a, b).

In their original form potentiometers have commonly been used to control electrical devices (for

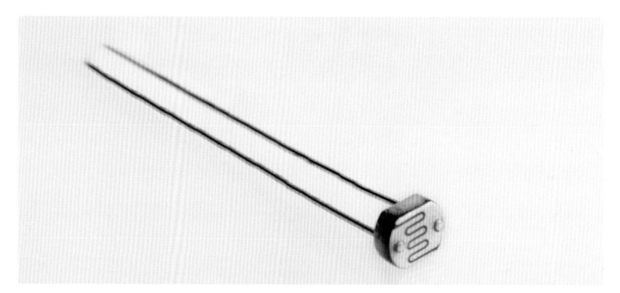

Figure 2.8 A light-dependent resistor is also called a photoresistor or a photocell; it is a light-controlled variable resistor.

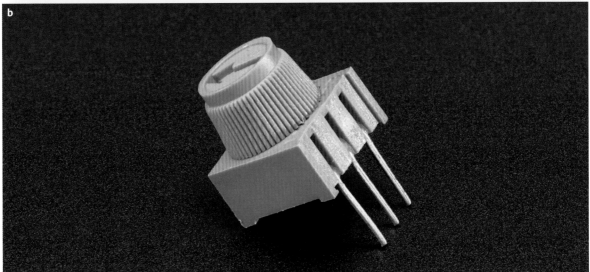

Figure 2.9 a, b Standard linear taper 1K ohm potentiometer with (a) metal knob and (b) colorful plastic knob. 10K trim potentiometers from Adafruit Industries have long grippy adjustment knobs and are perfect for breadboarding and prototyping.

example, as volume controls on audio equipment), but in fashion and tech projects they can be constructed from fabric and conductive and resistive materials so they can be included in a garment or a textile project (Figure 2.11).

LEDs

LED stands for *light-emitting diode*. It is a special type of diode (Figure 2.13) that lights up when electricity passes through it. Diodes are polarized

Basic Components of a Circuit

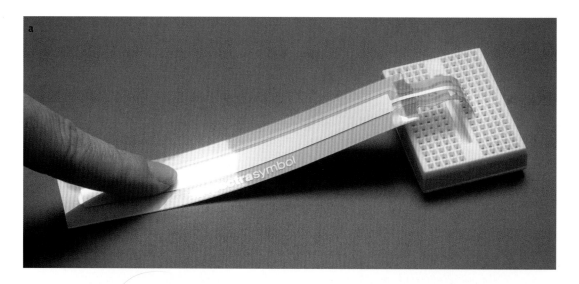

Figure 2.10 a, b (a) Linear soft potentiometer and (b) circular soft potentiometer. These are also known as Ribbon Sensors with adhesive backing from Adafruit Industries (www.adafruit.com). The resistance changes depending on where on the strip one presses.

and electricity is only intended to pass through in one direction. There are different indicators to let you know what direction electricity will pass through an LED, depending on the type of LED you are using. An electric bulb-like LED with legs will have a longer positive lead (anode) and a shorter ground lead (cathode). A sewable or a flat LED will have clear markings for positive and

Figure 2.11 This potentiometer is made by Kobakant, using the zigzag stitch on a sewing machine to embroider conductive and resistive traces side by side. A conductive object (spoon) can be used to bridge the contact between the traces and measure the change in resistance throughout the embroidered potentiometer.

Figure 2.12 This embroidered potentiometer by Paola Guimerans has a yarn wiper that can be moved along the conductive zigzag stitching to vary the resistance.

Basic Components of a Circuit   41

Figure 2.13 Various colors of sewable LEDs from Lilypad. The stripe on the back shows the color of the LED. (The first two LEDs on the left show purple and blue.)

negative leads, marked with a plus and a minus sign, respectively.

LEDs create a voltage drop in the circuit, but usually do not add much resistance. In order to prevent the circuit from shorting, it is a good practice to add a resistor in series. There are online LED/resistor calculators that can help you figure out how much resistance is needed for a single or multiple LEDs.

If you wire LEDs in series, keep in mind that each consecutive LED will result in a voltage drop until finally there is not enough power left to keep them lit. As such, it is ideal to light up multiple LEDs by wiring them in parallel. In order for that to work, make sure that all LEDs have the same power rating. Different colors may have different ratings, and each LED can only handle a limited amount of current and voltage. The specifications for each type of LED are covered in its datasheet, but for most common LEDs running at 5 volts, a resistor between 220 and 1K ohms will do the job.

Switches

A **switch** creates a physical break in a circuit, which prevents the flow of electricity (Figure 2.14). When you activate the switch, it opens or closes the circuit. It can be a mechanical device, such as a traditional light switch, or it can be activated through intentional input or a sensor. Complicated switches can open one or more connections and close others when activated. In fashion you don't

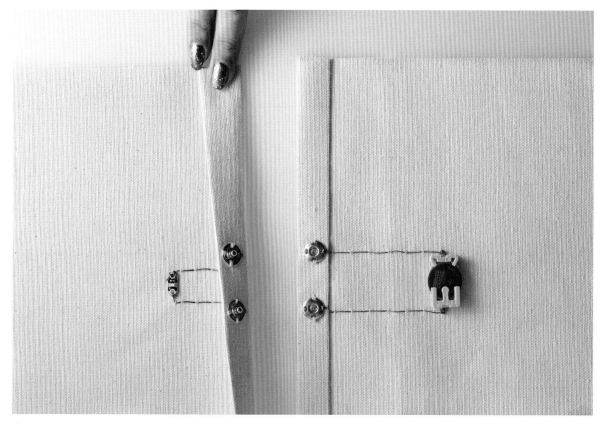

Figure 2.14 Metal snap switch. Both snaps need to be closed in order to complete the circuit.

have to limit yourself to mechanical buttons and levers. You can use hardware for switches, such as a zipper, a metal snap, a metal button, or a button (and buttonhole) covered with conductive thread or conductive fabric.

There are two types of switches: momentary and maintained. **Momentary switches** stay in a particular state only as long as they are activated, and they return to their original state when released (Figure 2.15). Their default state can be open or closed. A switch with a default-open switch will close the circuit when activated, and a switch with a default-closed state will deactivate the circuit when the switch is activated. **Maintained switches** retain their state until they are actuated into a new one. The switch physically keeps the circuit in a closed or open position. For example, the zipper pull in Figure 2.16 activates the switch once it is positioned past the conductive leads.

Even though there is a wide array of switches available off the shelf, they tend to be bulky and not well suited for a garment and most accessories. It is best to create your own switch based on available and desirable components. Take a look at the following case study of a flip knitted switch.

Please refer to the tutorials at the end of this chapter to create your own switch based on your collection concept.

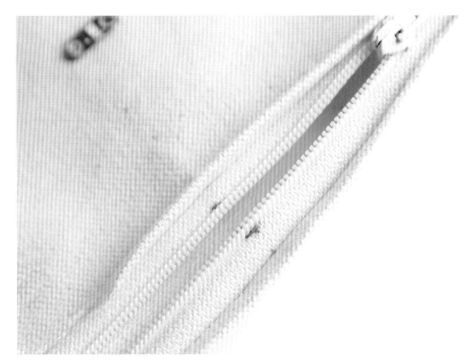

Figure 2.15 This momentary switch (zipper with plastic teeth and a metal zipper head) is currently off. The switch will be activated only at the moment when the head is positioned at the conductive thread leads. In any other position the switch is off.

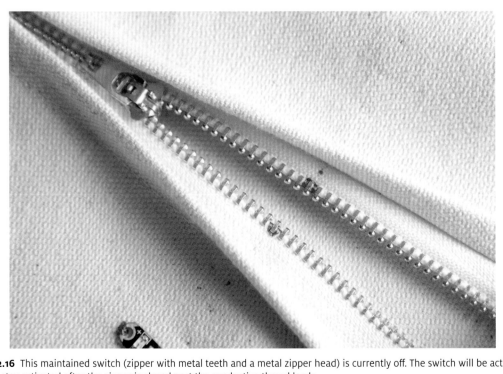

Figure 2.16 This maintained switch (zipper with metal teeth and a metal zipper head) is currently off. The switch will be activated and will stay activated after the zipper is closed past the conductive thread leads.

CASE STUDY
FLIP SWITCH

The following project was developed by Katharina Bredies, Hannah Perner Wilson, and Sara Diaz Rodriguez. This is a simple and creative way to make a flip switch with a knit yarn. The switch has two conductive stripes on either side of a protruding conductive flap, which can be moved to have contact with one or the other stripe. Each of the conductive stripes is connected to the positive side of two LED lights: one red and one green. The conductive flap is connected to the positive lead of the battery so that when the flap is open the battery power does not reach the LEDs, but when the flap makes contact with either stripe, power flows and the corresponding LED lights up. If the flap is pressed to be wide and flat and make contact with both stripes at once, then both LEDs light up.

The base of the switch is knit with non-conductive yarn and has two conductive stripes knit from Karl-Grimm thick silver conductive thread. The circuit connects the conductive stripes to the positive sides of two **SMD** (surface-mount device) LED lights: one red, one green.

The switch is controlled by an extended conductive flap, connected to the positive lead of the battery. When the flap is extended and does not touch either stripe, the circuit is deactivated and electricity does not flow from the battery to the LEDs.

The swatch is powered by a 3.7V LiPo battery encapsulated in hot glue and placed on top of a laser cut wood base with holes for the negative and positive sides to connect to the magnets of the circuit.

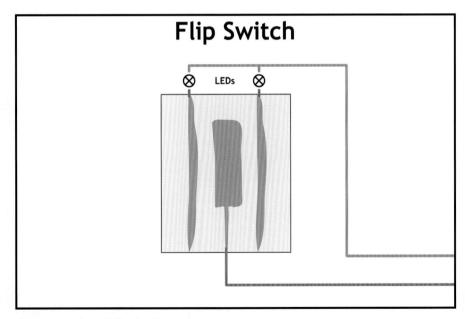

Figure 2.17 Schematic for the flip switch.

Basic Components of a Circuit

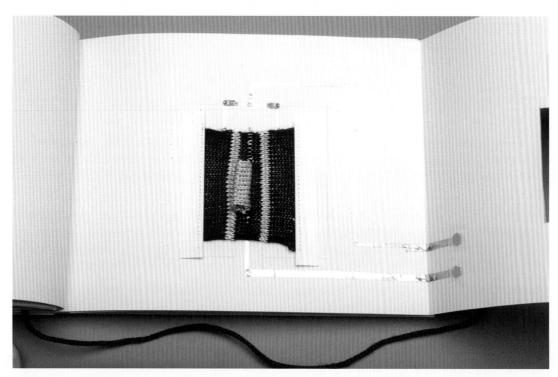

Figure 2.18 The full circuit is realized with conductive tape and presented in a swatch book.

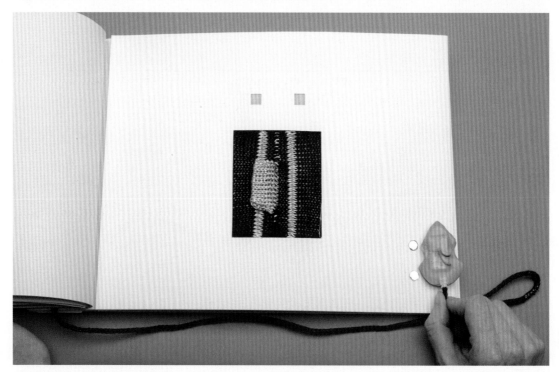

Figure 2.19 The flip switch battery.

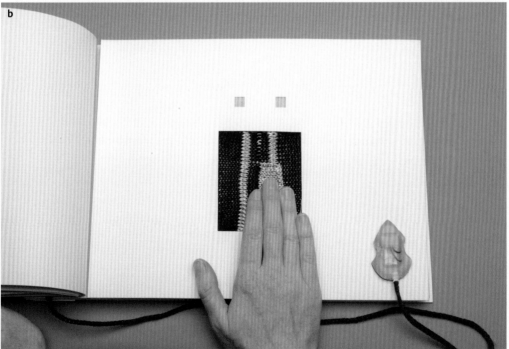

Figure 2.20 a, b When the flap is turned left or right and makes contact with either stripe, power flows and the corresponding LED lights up.

Basic Components of a Circuit

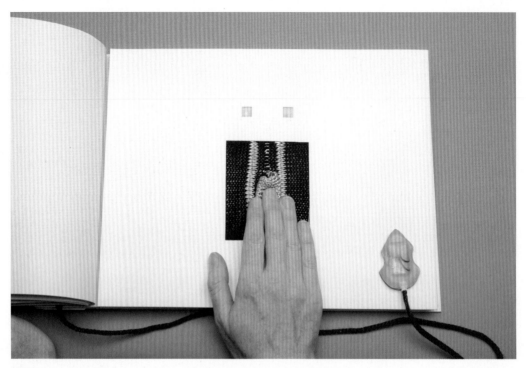

Figure 2.21 If the flap is pressed to be wide and flat and make contact with both stripes at once, then both LED lights light up.

INTERVIEW
KATHARINA BREDIES (CO-CREATOR OF THE FLIP SWITCH)

Katharina Bredies was born in Bremen, Germany, where she studied Integrated Design at the Hochschule für Künste. While in college she worked as a product designer, 3D modeler, and animator. She graduated in 2006 with a thesis based on the use of cybernetic systems analysis in design. Her interests cover research into interaction design and design theory as well as illustration, animation, and comic books.

She joined the Deutsche Telekom Laboratories in 2006 and worked as a research scientist and a PhD student. In her doctoral thesis, she investigated the value of irritation in design, and her main field of research covers e-textiles and the combination of traditional textile production techniques with electronic functionality.

AG, KM: You have worked as a product designer, 3D modeler, and animator? How did you transition into the world of electronic textiles and the combination of traditional textile production techniques with electronic functionality?

KB: I was trained as a product and interface designer during my studies. When I started working as a research scientist at Deutsche Telekom Laboratories [a Berlin-based research institute with a focus on information and communication technologies] in 2007, I also had the opportunity to start a practice-based PhD thesis in design research. I was interested in experimental tangible interface design. At the time I started to plan my final prototypes for the thesis in 2010, the Arduino Lilypad and conductive threads were still quite new. It made a lot of sense to me then to experiment with textiles as an interactive medium in my thesis because it was so obviously different from conventional electronics. This is how I also became interested in traditional textile production techniques. There is such an interesting contrast between the old technologies of weaving and knitting, and new ones like electronics—what we perceive as "technology" nowadays. It is an interesting challenge to explore the interaction potential of textile materials, to use their softness as a quality for tangible interactions with computers.

AG, KM: What leads your personal creative process? Do you start with electronics you would like to work with, or do you start with a concept and then look for the tech that would help that vision?

KB: When I develop textile sensors and switches, I often start with a concept, usually an idea of an interaction that is appropriate for textile materials—for example folding, crumpling, or knotting. Such typical actions that one can perform with fabric need a more specific shape on which they can be performed, so I do a couple of sketches to specify the form of the interactive object. In this stage, I also start to think about how the object could be constructed with conductive threads or fabrics, and a specific production technique. It would of course be perfect if I could just construct the whole circuit with all components from fiber materials, but this is not really possible now. So once I know what the sensor and the circuit should look like, I get the other electronic components that I need. The starting point for the design is therefore between the electronics and the textile, ideally a synthesis of both.

AG, KM: You are working within the state-of-the-art area of electronics and interactive textiles. Do you consider yourself a designer first or a tech innovator?

KB: I consider myself a designer. Technological innovation is a specific form of design to me. Depending on who looks at the results, it might be interpreted as tech innovation as well.

AG, KM: Developing wearable technology is one of the leading innovative industries nowadays with giant tech industries leading the way. What do you think are the biggest challenges in designing wearable technology?

KB: One obvious challenge for the industry in bringing wearables into the market is probably the robustness of their products, and how well they survive the washing machine. But the bigger challenges to me are to adopt textile production techniques and materials for electronics on the one hand, and to look beyond the textile shapes that we are familiar with nowadays on the other. A lot of wearable products are not textile, and depending on their purpose, they don't have to be. It might turn out that wearables are actually not the best application of electronic textiles in the first place, because they have such high requirements when it comes to comfort and long-term durability.

AG, KM: What innovations in e-textiles and new materials would you like to see happen in the race for inventive and functional wearables?

KB: There is actually a lot of interesting research in the area of fiber power supplies and semiconductors. It would be great to be able to purchase and work with those materials. I would like to be able to knit or weave my batteries and transistors.

AG, KM: What are your thoughts on connecting our garments, our accessories, maybe even our bodies to the Internet twenty-four-seven?

KB: We will probably see this happen in the next years, as the discussion about the Internet of things is already ongoing for some time now. For me it is a matter of design—how we handle the data that we produce and share—and that should never be done carelessly. This is important to remember because sometimes those man-made developments are treated like natural phenomena. We will probably have a huge debate in the future on the legislation regarding personal data. For me it is very ambiguous to contribute to the data accumulation, and I personally prefer to keep users in control of their data.

AG, KM: What advice do you have for young fashion designers who might be hesitant to enter the world of wearable tech or integrate technology into fashion?

KB: The best thing to do is to seek contact with people who work in the field, then watch and learn. If there is a one-day workshop in your hometown or as part of a conference, it is a very good way to get first hands-on experience. There are by now a lot of hardware parts that are readymade and easy to use, and a lot of support from the online community. It is, however, important to talk to people with a similar background, who remember how confusing it can be to work with wearable tech in the beginning.

TUTORIALS

In this chapter you were introduced to the basic principles of electricity and acquired the knowledge to build an electric circuit with various electronic components. The following soft circuit tutorials will reveal the step-by-step techniques you can use to build your own circuits.

These tutorials provide simple but powerful techniques. You should design and execute your circuits based on your own concept design ideas. Each circuit can be shaped to fit your personal aesthetic; you can use materials based on your own color palette and desired look. Surface treatment can be used to either hide or reveal parts of the conductive leads, thread, or the actual LEDs. Don't forget: You are the designer; electronics are simply a tool to help you realize your design ideas.

TUTORIAL 1:
BASIC SOFT CIRCUIT WITH VISIBLE AND INVISIBLE STITCH LINE

In this tutorial we create two variations of a basic soft circuit with one sewable LED. One variation clearly shows the conductive thread on the surface of the fabric, while the other shows you an option of hiding the conductive thread from the surface through a blind stitch.

This tutorial gives you the basic skills to create your own circuit, but should not restrict what textiles you use and what the shape of your own circuit is. You should create the circuit that you need depending on your design. For example, with this particular case we have outlined a rectangular shape for the conductive thread leads, but you can make a circular shape and place the LED closer or further away from the battery. Whatever shape you choose, make sure that you create a complete and closed circuit.

Materials List

Fabric, pencil, conductive thread, coin cell battery, battery holder, LED, needle. Yours may vary, depending on your choice and size of fabric, as well as drawing tools.

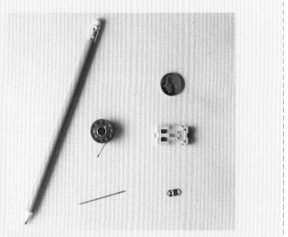

Figure 2.22 Materials.

Basic Components of a Circuit

Step 1 (Figure 2.23): Position the battery holder and the LED according to your design and outline the leads for your conductive thread with a pencil or chalk. Make sure your marks are removable if you mark the right side of the fabric. Be careful to orient the battery holder and the sewable LED so that the negative lead of one is connecting to the negative side of the other. You will see clear − and + marks on both components.

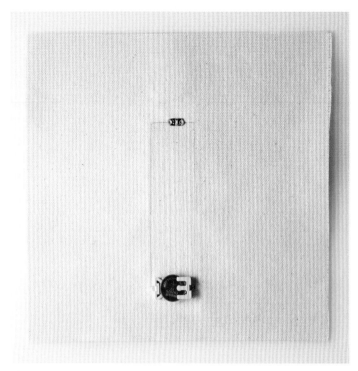

Figure 2.23 Step 1.

Step 2 (Figure 2.24): Starting with the battery holder, secure the negative lead by sewing through the hole of the conductive part, and then stitch along the pencil mark towards the negative side of the LED.

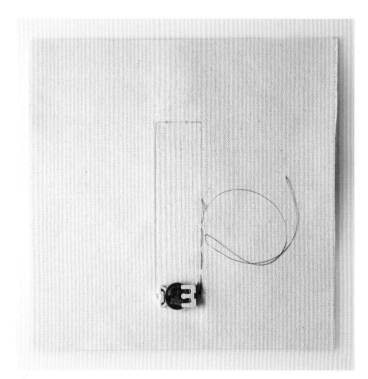

Figure 2.24 Step 2.

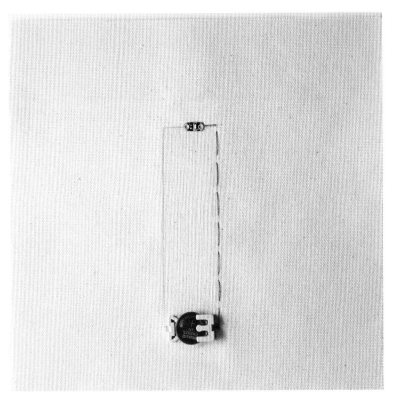

Figure 2.25 Step 3.

Step 3 (Figure 2.25): Continue the stitch line all the way to the negative side of the LED. Then finish by looping the thread a few times through the LED lead.

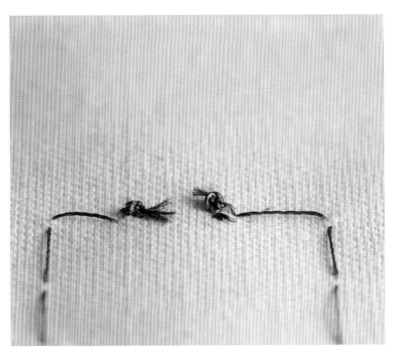

Figure 2.26 Step 4.

Step 4 (Figure 2.26): Cut and tie the thread at the LED lead on the wrong side of the fabric. You can secure the thread knot with fray check or glue. Start a new thread stitch on the other side of the LED towards the battery along the outlined path.

Step 5 (Figure 2.27): Finish by tying the conductive thread to the conductive lead of the battery holder on the wrong side of the fabric. The finished soft circuit displays visible conductive thread on the right side of the fabric.

Figure 2.27 Step 5.

Step 6 (Figure 2.28): Make sure that each end of the LED and the battery holder is secured by knotting the conductive thread, and then finish with a dot of glue or fray check on the back of the circuit. There should not be any long thread ends. Loose threads can come into contact with each other and short the circuit.

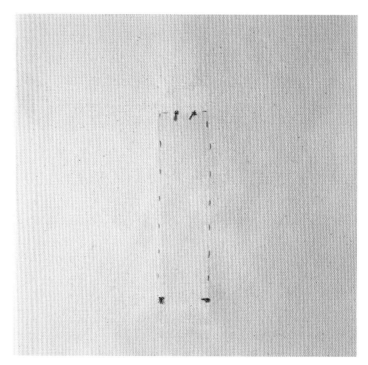

Figure 2.28 Step 6: Back view of completed soft circuit.

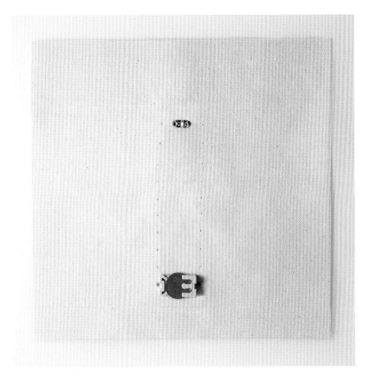

Figure 2.29 Blind stitch.

The same circuit can be recreated with a blind stitch. We used the same exact materials and the same exact canvas fabric, but we stitched with a blind stitch to show the option of hiding the conductive thread on the right side of the fabric (Figure 2.29). Most of the time you will not want to show the conductive thread on the surface in order to preserve the integrity of your design. You can use this method to hide your stitches or use a more decorative stitch on the surface to better suit your own design aesthetic. The choice is yours. The back view of the soft circuit with blind stitch (Figure 2.30) shows that the majority of the conductive thread for this circuit is visible on the wrong side of the fabric.

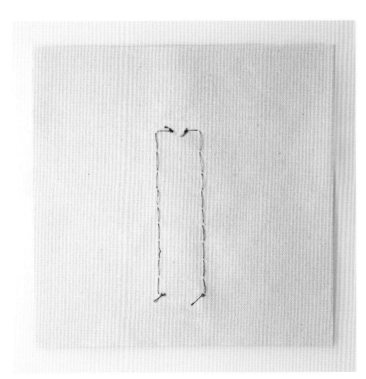

Figure 2.30 Blind stitch, back view.

Basic Components of a Circuit

TUTORIAL 2:

SOFT CIRCUIT WITH LEDs CONNECTED IN PARALLEL VS. SERIES

This tutorial demonstrates how to build one soft circuit with LEDs connected in parallel and another with LEDs connected in series.

For a circuit with LEDs connected in parallel, start by connecting one LED to the power source and then keep adding as many LEDs as you need, but keep in mind that you are limited by the electrical characteristics of the battery.

Materials List

Fabric, pencil, conductive thread, coin cell battery, battery holder, 3 LEDs, needle. Yours may vary depending on your choice of fabric, drawing tools, and the number of LEDs you want to use.

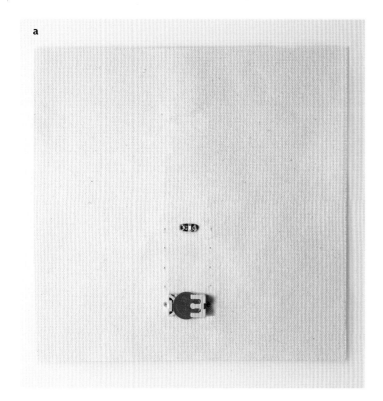

Step 1 (Figure 2.31 a, b): Using the stitching technique demonstrated in Tutorial 1, connect the first LED to a coin cell battery with blind stitches so that the soft circuit is invisible on the right side of the fabric. The positive side of the LED should connect to the positive side of the battery holder and, respectively, the negative to the negative.

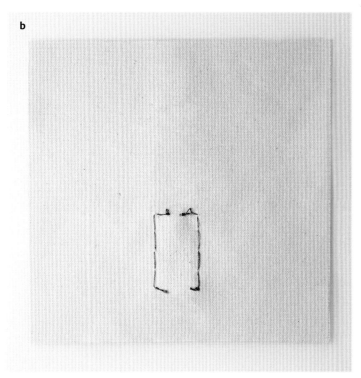

Figure 2.31 a, b Step 1.

Step 2 (Figure 2.32 a, b): Add a second LED in parallel by connecting the negative end of the first LED to the negative lead of the second and the positive to the positive. You can achieve that by carefully knotting a new conductive thread to the existing stitch line leading from the battery to the lead of the first LED and then continuing the stitch line towards the second LED.

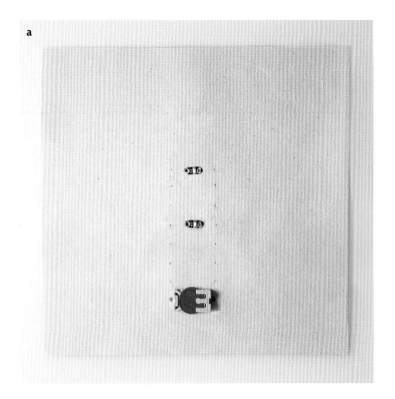

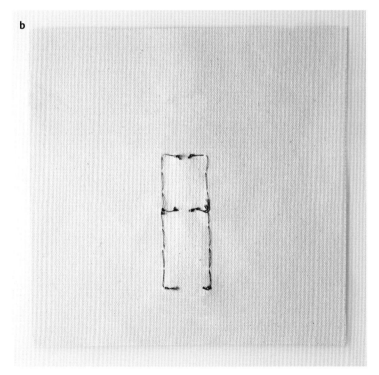

Figure 2.32 a, b Step 2.

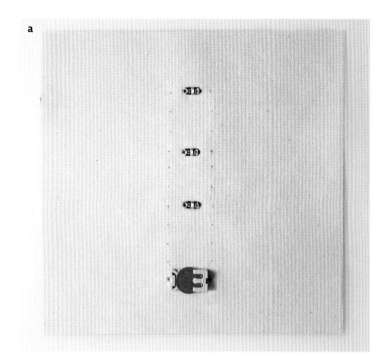

Step 3 (Figure 2.33 a, b): Using the same method, add a third LED to create this parallel soft circuit. When finished this circuit demonstrates that you can have multiple LEDs within the same circuit powered by the same battery as long as they are connected in parallel.

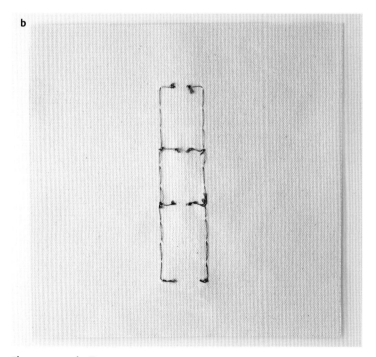

Figure 2.33 a, b Step 3.

Basic Components of a Circuit

Using the same supplies and sewing techniques you can connect the LEDs in series, but you must increase the power supply or there won't be enough power to light up your circuit. The coin cell battery will power one LED, but adding additional LEDs connected in series will result in failure (Figure 2.34).

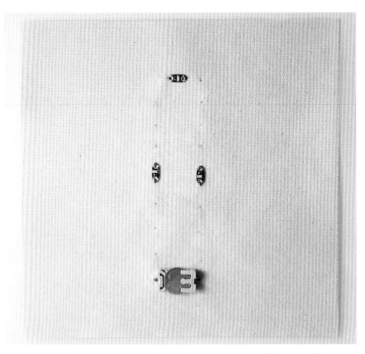

Figure 2.34 Failure: Using a coin cell battery does not provide enough power to light up multiple LEDs. You must increase the energy source to power up this circuit.

TUTORIAL 3:

MOMENTARY SWITCH (PLASTIC ZIPPER) AND MAINTAINED SWITCH (METAL ZIPPER) CIRCUITS FOR A SHORT ZIPPER

In some garments or accessories you will have to work with a short zipper. A good example of that is a sleeve opening at the wrist, or zipped pocket of a garment or accessory. In these cases the soft circuit can be sewn up and above the end of the zipper. We will look at two different options: a momentary switch, which can be created with a plastic-teeth zipper, and a maintained switch, which can be made with a metal-teeth zipper. This tutorial demonstrates the construction for these two options with 6" zippers.

Materials List

Materials list for momentary switch: fabric about 10"x 20", pencil, conductive thread, coin cell battery, battery holder, 1 LED, needle, 6" plastic zipper. Yours may vary depending on your choice of fabric, zipper length, and drawing tools.

Materials list for maintained switch: fabric about 10" x 20", pencil, conductive thread, coin cell battery, battery holder, 1 LED, needle, 6" metal zipper. Yours may vary depending on your choice of fabric, zipper length, and drawing tools.

Step 1: To prepare for the tutorial, attach a zipper to the short sides of the fabric and close the remainder of the seam to create a closed loop, as shown in Figure 2.35. There should be about 4" above the end of the zipper to the edge of the fabric. That will allow the conductive thread to close the connection and create a closed circuit. Repeat the same process for the second set of fabric and zipper. If the zipper head of the momentary switch is painted, you need to file away some of the paint on the bottom and sides in order to reveal the metal at the contact points and create a solid connection. Keep in mind that paint will prevent conductivity and will not allow electricity to flow through the zipper head.

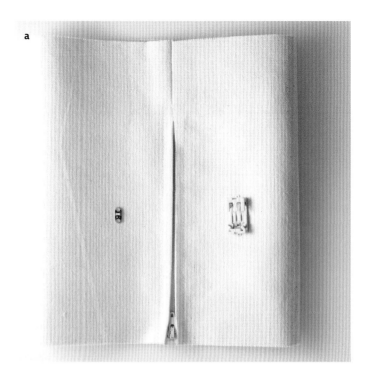

Step 2: (Figure 2.35 a, b) Position the LED and the battery case and mark the path for the soft circuit on the wrong side of the fabric for both zippers. Make sure the positive side of the LED matches with the positive side of the battery case.

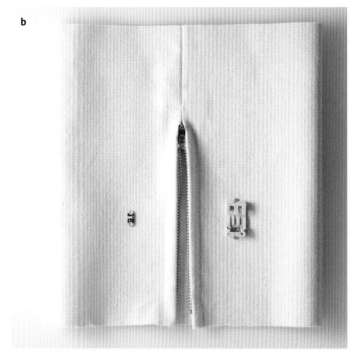

Figure 2.35 a, b Step 2 (a) Momentary switch (zipper with plastic teeth); (b) maintained switch (zipper with metal teeth).

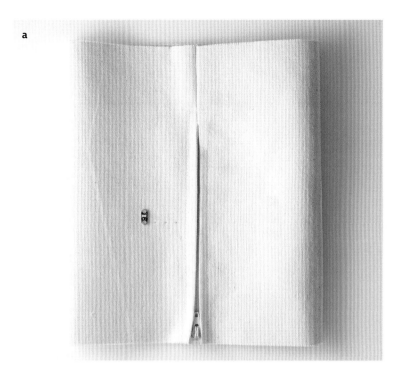

Step 3 (Figure 2.36 a, b): Sew the LED with a blind stitch from the LED lead to the zipper teeth. Loop the thread a few times in between the zipper teeth to create a better connection. Repeat the same process for the second swatch. We chose to place the LEDs on the outside and the battery cases hidden on the inside for both of these samples. You can choose to hide the LED on the inside or stitch the thread on the right side of the fabric.

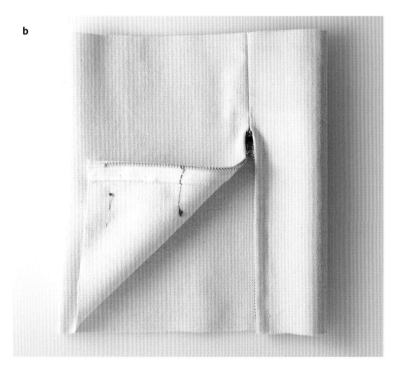

Figure 2.36 a, b Step 3 (a) Momentary switch (zipper with plastic teeth); (b) maintained switch (zipper with metal teeth).

Basic Components of a Circuit

Step 4 (Figure 2.37 a, b): Stitch the battery case with conductive thread on the opposite side of the zipper on the inside of the swatch. Then carefully guide along your thread stitches to the same exact spot of the zipper teeth to match the opposite lead from the LED. When you reach the zipper teeth, loop the thread a few times to create a solid conductive connection. Repeat for the other swatch.

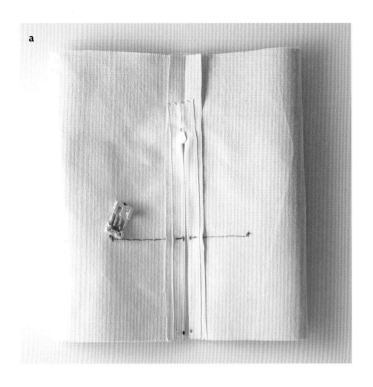

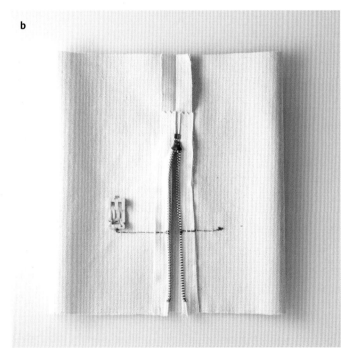

Figure 2.37 a, b Step 4 (a) Momentary switch (zipper with plastic teeth); (b) maintained switch (zipper with metal teeth).

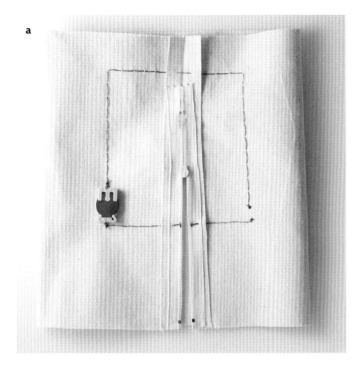

Step 5 (Figure 2.38 a, b): Continue to sew the circuit from the battery case up and above the zipper to the opposite lead of the LED. This step completes the circuit. Repeat for the other swatch.

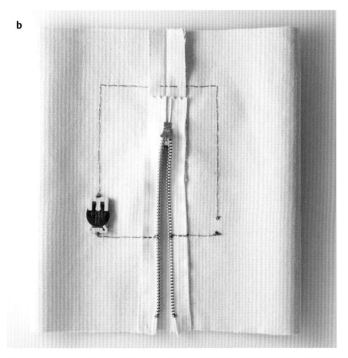

Figure 2.38 a, b Step 5 (a) Momentary switch (zipper with plastic teeth); (b) maintained switch (zipper with metal teeth).

When the zippers are open, the switches are off for both circuits and the LEDs do not light up (Figure 2.39).

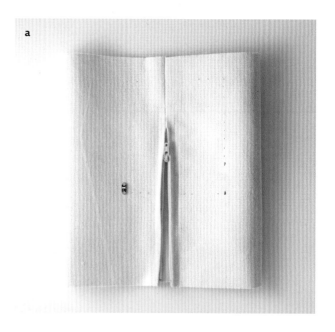

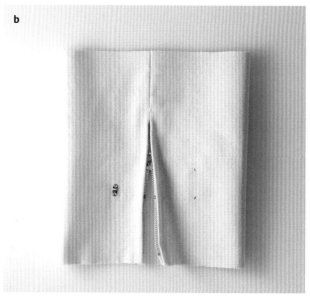

Figure 2.39 a, b (a) Momentary switch (zipper with plastic teeth); (b) maintained switch (zipper with metal teeth).

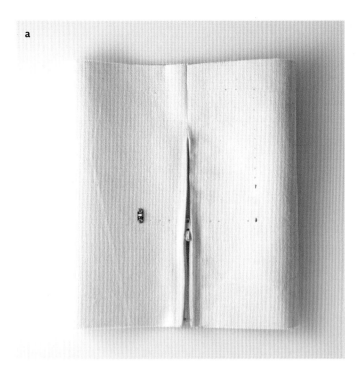

In order to activate the momentary switch circuit, you must position the zipper head exactly at the point where the conductive thread from the battery case meets with the conductive thread from the LED. At that—and only at that—point the switch is on and the LED lights up (Figure 2.40 a). Once the zipper pull is positioned past the contact point, the switch turns off and the LED goes off (Figure 2.40 b).

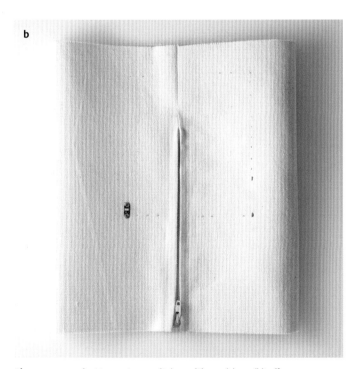

Figure 2.40 a, b Momentary switch positions: (a) on (b) off.

In order to activate the maintained switch circuit, you must position the zipper head anywhere past the point where the conductive thread from the battery case meets with the conductive thread from the LED. The switch is on past the contact point and the LED lights up. The switch stays on as long as the zipper is closed (Figure 2.41).

Figure 2.41 Maintained switch: on.

TUTORIAL 4:

MOMENTARY SWITCH (PLASTIC ZIPPER) AND MAINTAINED SWITCH (METAL ZIPPER) CIRCUITS FOR A LONG ZIPPER

In some garments or accessories you have to work with a long zipper. To complete a circuit in this scenario presents a unique challenge. For example, a center front zipper that opens a jacket does not allow for a soft circuit to go up and above the zipper. In such a case you have to build a switch that goes all around the full width of the jacket. The same can be used in a sleeve on which you want to go around the arm if the zipper is too long. A similar scenario may be needed in a bag or other accessory.

For this tutorial we look at two different options: a momentary switch, which can be created with a plastic-teeth zipper, and a maintained switch, which can be made with a metal-teeth zipper. This tutorial demonstrates the construction for these two options.

Materials List

Materials list for the momentary switch: fabric 10"x20", pencil, conductive thread, coin cell battery, battery holder, 1 LED, needle, 10" long plastic zipper to cover the length of the fabric. Yours may vary, depending on your choice of fabric, zipper length, or drawing tools.

Materials list for the maintained switch: fabric 10"x20", pencil, conductive thread, coin cell battery, battery holder, 1 LED, needle, 10" long metal zipper to cover the length of the fabric. Yours may vary, depending on your choice of fabric, zipper length, and drawing tools.

Step 1: To prepare for the tutorial, attach the zipper to the short sides of the fabric. The closed end of the zipper should reach to the edge of the fabric. Once you zip the sample closed, it should close the fabric into a loop as shown in Figure 2.42.

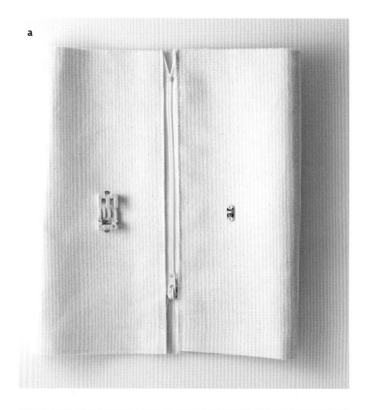

Step 2 (Figure 2.42 a, b): Position the LED and the battery case and mark the path for the soft circuit on the wrong side of the fabric for both zippers. Make sure the positive side of the LED matches with the positive side of the battery case.

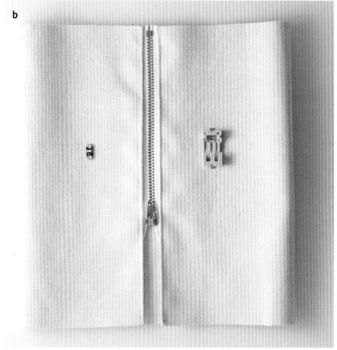

Figure 2.42 a, b Step 2 (a) Momentary switch (zipper with plastic teeth); (b) maintained switch (zipper with metal teeth).

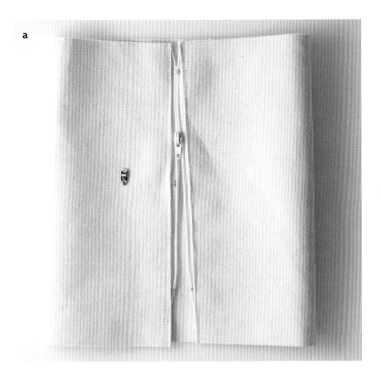

Step 3 (Figure 2.43 a, b): Sew the LED with a blind stitch, starting from the LED lead and continuing to the zipper teeth. Loop the thread a few times in between the zipper teeth to create a better connection. Repeat the same process for the second swatch. Repeat the same process for the maintained switch with metal zipper. For this particular project we chose to place the LEDs on the outside of the swatch and the battery cases hidden on the inside. You can position yours on either side.

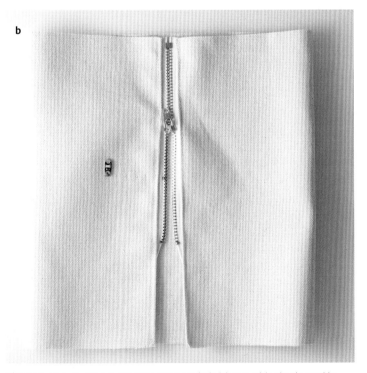

Figure 2.43 a, b Step 3 (a) Momentary switch (zipper with plastic teeth); (b) maintained switch (zipper with metal teeth).

Step 4 (Figure 2.44 a, b): Flip the swatch on the inside and stitch the battery case with the conductive thread. Carefully guide the stitch line to the same exact spot of the zipper teeth to match the opposite lead from the LED and loop the thread a few times around the zipper teeth to create a better connection. Repeat the same process for the other swatch.

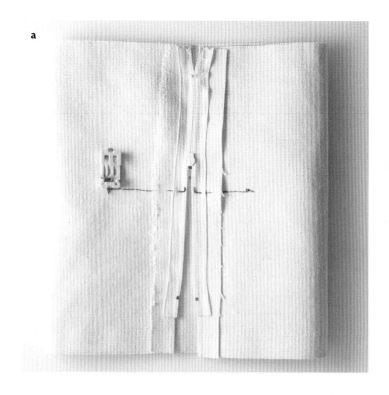

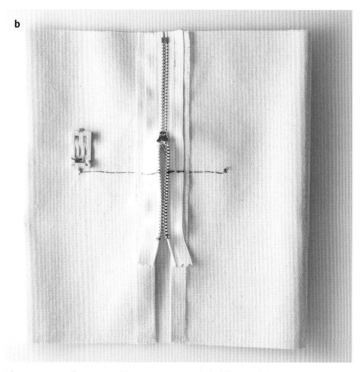

Figure 2.44 a, b Step 4 (a) Momentary switch (zipper with plastic teeth); (b) maintained switch (zipper with metal teeth).

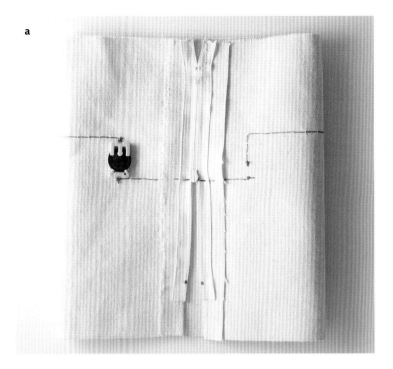

Step 5 (Figure 2.45 a, b): Continue to sew the stitch line of the circuit from the opposite side of the battery case around to the opposite lead of the LED. This completes the circuit. Repeat for the other swatch.

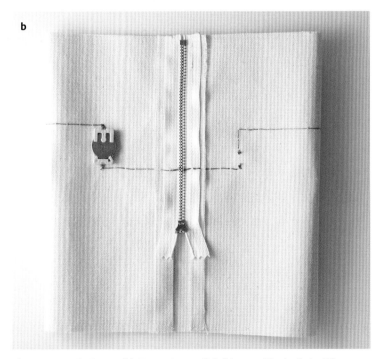

Figure 2.45 a, b Step 5 (a) Momentary switch (zipper with plastic teeth); (b) maintained switch (zipper with metal teeth).

Keep in mind that a line of a single conductive thread will go around the full width of the swatch regardless of the kind of switch you have (Figure 2.46). If you do not like to see the thread, make sure you use a blind stitch or conceal it with a surface texture of your design. Naturally you can change the shape of the stitch line and create a design that works better for you instead of using straight lines and sharp angles. Keep in mind that the longer and more complicated the circuit is, the more possibilities you have of breaking it.

When the zippers are open, the switches are off for both circuits and the LEDs do not light up (Figure 2.47 a, b).

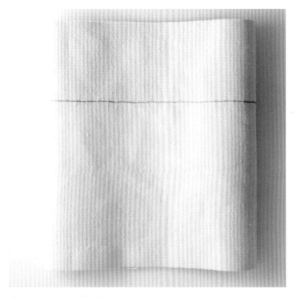

Figure 2.46 Full width.

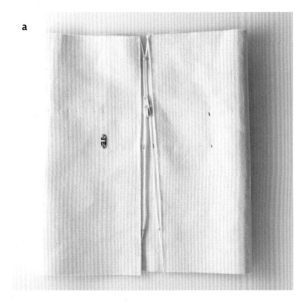

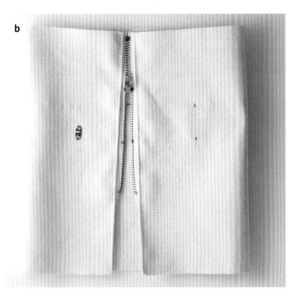

Figure 2.47 a, b (a) Momentary switch (zipper with plastic teeth); (b) maintained switch (zipper with metal teeth).

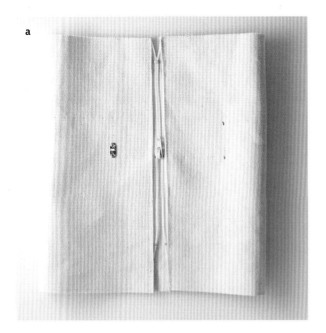

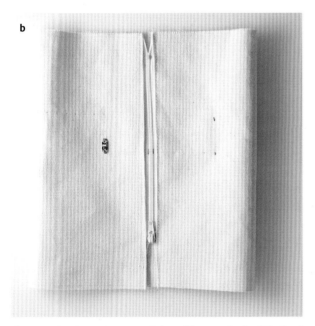

Figure 2.48 a, b Momentary switch positions: on (top), off (bottom).

In order to activate the momentary switch circuit, you must position the zipper head exactly at the point where the conductive thread from the battery case meets with the conductive thread from the LED at the zipper teeth. At that—and only at that—point the switch is on and the LED lights up (Figure 2.48 a). Once the zipper pull is positioned past the contact point, the switch turns off and the LED goes off (Figure 2.48 b).

In order to activate the maintained switch circuit, you must position the zipper head anywhere past the point where the conductive thread from the battery case meets with the conductive thread from the LED. The switch is on past the contact point and the LED lights up. The switch stays on as long as the zipper is closed (Figure 2.49). In this case the zipper head was painted white, and we had to file away some of the paint in order to reveal the metal and establish a connection at the contact points.

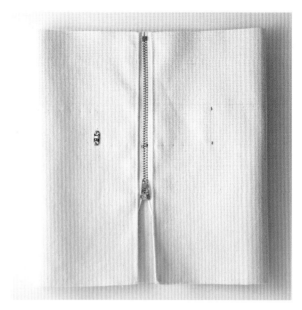

Figure 2.49 Maintained switch: on.

INTERVIEW
KOBAKANT

Mika Satomi and Hannah Perner-Wilson, who have been collaborating since 2006, formed the collective KOBAKANT in 2008. As stated on their website (www.Kobakant.at): "Through their work, they explore the use of textile crafts and electronics as a medium for commenting on technological aspects of today's 'high-tech' society. KOBAKANT believes in the spirit of humoring technology, often presenting their work as a twisted criticism of the stereotypes surrounding textile craftsmanship and electrical engineering. KOBAKANT believes that technology exists to be hacked, handmade and modified by everyone to better fit our personal needs and desires."

Hannah and Mika have been on the forefront of exploring soft circuits and textiles, as well as exploring techniques, developing tutorials, and holding workshops around the world.

AG, KM: Hannah, your background and education is in both Industrial Design and Media Arts and Sciences. What lead you to the fields of electronic textiles and wearable computing?

H: During my industrial design studies I was introduced to a whole range of model-making materials

Figure 3.50 Mika Satomi (left) and Hannah Perner-Wilson (right).

for producing mock-up prototypes as design sketches. After taking a course on sensor technologies I was struck and inspired by the fact that I could build my own electronic devices. I enjoyed building things that were not just mock-ups but functioning electronic items, and the fact that I could build these items from materials that were not the standard electronic building blocks, but instead "raw" materials like copper tape, carbon foam, conductive fabric and resistive yarns, meant I had to solve details of material connections and their integration with other parts. This challenge to solve such crafty issues in order to be able to invent and build unusual technology is what drew me in. After completing my bachelor's degree in Austria, I had the luxury of taking a year to further explore this practice, setting up a studio workspace in my apartment where I experimented and made things. Encounters with everyday computer technology mixed with all the unusual conductive materials I can get my hands on still inspire much of my work today.

AG, KM: Mika, your education is in Graphic Design and Media Creation and you were a researcher at The Smart Textile Design Lab at Textilehögskolan in Borås, Sweden. Please tell us how and why you entered the world of e-textile crafts, soft circuits, and wearable tech?

M: My first encounter with the idea of wearable technology was when I took a course in 2006 at the Art University Linz Austria. Later that year, Hannah and I started to work on a project called "Massage Me" in which we embedded a PlayStation game controller on the back of a jacket. During the development of this project, we encountered conductive textile materials, and from there we started to work with this material and explore methods for creating soft circuitry. From the spring to late summer of 2009, we had a research position at the Distance Lab, Scotland, where we created a database for soft circuitry materials and technique called "How To Get What You Want." During this time, we had the luxury to spend time on experimenting on the various materials, tools, and techniques and construct the basic structure of this database. We are still constantly updating this database, and it was a big part of our work until now. After I finished my term at the Distance Lab, I had the good luck to work with trained textile designers at the Smart Textile Design Lab on the topic of Smart Textiles. As it is located in a textile school, we particularly looked at the materials and textile techniques that can be applied in e-textiles. This gave me a very good introduction to textile techniques such as weaving, knitting, printing, and dyeing. Also we had access to industrial textile machines in the school, which led us to explore different scales and methods. For example, Hannah visited us as a guest researcher and we worked on an industrial jacquard weaving machine and wove a series of conductive fabric. This experiment resulted in an e-textile collection called "Involving the Machines."

AG, KM: The two of you have been collaborating on textile crafts and electronics since 2006 and have started the now-famous collective KOBAKANT. How did the collaboration between the two of you start and how has it evolved over the years?

M: The first few years of our collaboration, we were not always living in the same city, and projects often happened in a short, concentrated production period, often as an artist in residency at some institute or while visiting each other for a few weeks' period to work on a project. Last year we started living in the same city and have an atelier together.

AG, KM: What is your philosophy and approach to working with textile crafts and electronics?

H: I enjoy exploring the less common route of doing things. It feels more adventurous to be doing things differently. But I also believe very rationally that there is tremendous value in creating diversity within any practice. I see "classical" electrical engineering practices as taking a route of optimization for compartmentalizing electronic functionality into "parts," making building electronics a very standardized and streamlined process. I want to create alternative and different ways of working with electricity that leverage other materials, techniques, and skills.

M: For me, it is "Do It Yourself," literally. I am a bit of a control freak, and I like doing the whole process myself. And in this sense making electronics from scratch, from textile to sensors to circuit, instead of buying readymade electronics, suits me well. We wrote in our bio that "KOBAKANT believes in the spirit of humoring technology," and perhaps you can say this is our philosophy.

AG, KM: You have developed and taught countless workshops around the world. What is your favorite and why?

H: The workshops in which we start by taking apart existing electronics are some of my favorite ones. The act of opening up a designed and functional object in order to understand its workings feels like an intuitive approach to understanding technology. It also encourages a more exploratory approach, learning-by-doing. And it motivates why creating electronics from textiles is a different approach to embedding functionality into the material, rather than packing it away inside black boxes.

M: The length of a workshop can vary from a few hours to a semester-long university course. Short workshops are interesting as you see the power of making emerging within a person in short period of time by simply encountering materials and techniques, but longer workshops, especially semester-long courses, are my favorite as you can follow the entire development of a prototyping phase with iterations. These longer courses can touch upon depth of the techniques and discourses around the topic, which I personally enjoy very much being a part of.

AG, KM: You are very involved with the DIY electronic craft and wearable tech community. What do you think is the appeal of this area of creative development and why are people attracted to it?

H: As with anything new, people are attracted to novelty, but for different reasons. Some are excited to get things working, solving technical, material, integration, and manufacturing issues; others are designing applications and speculating on a future where soft circuits and wearable technology is ubiquitous. What is empowering about approaching this field with a Do-It-Yourself perspective is that you are in a position to invent, imagine, and realize projects in this space all by yourself or with others. Your projects don't have to appeal to anybody but you. I see the DIY community as a source of very personal creative expression and innovation.

AG, KM: You have created an extensive amount of custom-made sensors and controllers. What's the value provided of handcrafted sensors vs. conventional manufactured components?

H: Good question. Manufactured components were also once handmade, but they transitioned through a process of optimization and standardization to become building blocks. For me the value in making components, or "parts," yourself lies in building without blocks. You have control over the shape, size, color, and materiality of even the smallest parts of your design. If you are more flexible the results you produce will be unique.

M: When working on projects, you may want to or need to use specific shape, color, texture, or function that you design. Think of choosing a button for your garment: You may buy a button from a shop, but also you may make your own button to fit your design. As Hannah said, in design processes, you often need building blocks that are not standardized or not even existing. So, you need to make your own building blocks, or rather you build without blocks. The value of making your own sensors is to be able to achieve a freedom in pursuing your design without the limitation of available off-the-shelf components.

AG, KM: What technical innovations would you like to see in the world of soft circuits and electronics?

H: I would love to see an emphasis put on the materiality of electricity. Get away from thinking of a resistor as a part, but as a material that can be cut, cast, mixed, and carved into both a desired value and form. I also believe that this approach to electronics offers different ways of understanding and engaging with technology that will help add more diversity to the electronics we find ourselves surrounded by.

M: I would like to see developments on more specialized techniques and materials for e-textiles. At the moment, many components, techniques, and tools we use are adapted from conventional electronics practices or textile practices, which are somehow limiting. I wonder if we can come up with new ways, for example, to knit, to weave, or to drape for e-textiles practices. Will it take us to totally new aesthetic and functions we could not imagine now? This will be fascinating.

AG, KM: What advice do you have for fashion designers who might be hesitant to work with technology and electronics?

H: Take things apart. Start with a simple device like a bike light or a gadget that makes sound. Try replacing some of the parts you find inside the device with things you can make yourself from materials you're comfortable working with. Draw what you see and try to make sense of things through trial-and-error experiments. And don't give up easily. The patience to de-bug your project is an incredibly valuable skill. Oh, and document your work and share it with others!

M: Electronics objects do not "think" or "guess," unfortunately (or fortunately). When things do not work, there usually is a logical explanation. It does not simply feel like "not working" and it cannot guess what you mean. Things that look complicated are the accumulation of simple logics. How a fabric drapes down on a three-dimensional body is a far more complex system than many of the electronics circuits. The only difficulty is that we are not used to this system. Be patient and try to get accustomed to the logic of bits and bytes. You will get familiar with this system, as you are familiar with the flow of fabrics.

For Review

1. What is an electric current and what are the two types of electricity?
2. Name the basic components of a circuit.
3. What is the difference between a closed and an open circuit?
4. How are components wired in a parallel circuit versus series circuit?
5. What are the main benefits of using a flexible circuit?
6. How would you change the value of a fixed resistor?
7. What is the function of a switch?
8. What is the difference between momentary switches and maintained switches?

For Discussion

1. What traditional fashion components can be used successfully in a closed circuit on a garment?
2. Which ones will work well as switches and why?
3. Why is it important to integrate the function of closing and opening the circuit into the garment?
4. How can you power a soft circuit and how can the power source be integrated in the garment or accessory?

Books for Further Reading

Hartman, Kate. *Make: Wearable Electronics: Design, Prototype, and Wear Your Own Interactive Garments*. Sebastopol: Maker Media, 2014.

Pakhchyan, Syuzi. *Fashioning Technology: A DIY Intro to Smart Crafting*. Sebastopol, CA: Make, 2008.

O'Sullivan, Dan, and Tom Igoe. *Physical Computing: Sensing and Controlling the Physical World with Computers*. Boston: Thomson, 2004.

Online References

Bredies, Katharina, Design Research Lab, http://www.design-research-lab.org/persons/katharina-bredies/, Web. 17 June 2015.

Igoe, Tom, ITP Physical Computing, https://itp.nyu.edu/physcomp/, Web 17 June 2015.

Satomi, Mika, and Perner-Wilson, Hannah, KOBAKANT, http://www.kobakant.at/, Web. 17 June 2015.

Key Terms

AC and DC current: AC means that the current is alternating, and DC means that the current is direct.

Anode: The negative terminal in the cell of a battery.

Cathode: The positive terminal in the cell of a battery.

Circuit: A complete and closed pathway or a never-ending loop through which an electric current can flow uninterrupted.

Conductor: An object, material or fabric that permits the flow of electric charges in one or more directions.

Current: The amount of electrical energy passing through a particular point, rated in Amps (A) or Milliamps (mA).

Electrical resistance: An electrical conductor is the opposition or resistance to the flow of an electric current through that conductor.

Electric current: A flow of electric charge, traveling from a point of high electrical potential, identified as power or the symbol for plus (+), to the lowest, which is usually identified as ground or the symbol for minus (−).

Fixed resistor: A component that adds resistance to a circuit at a predetermined value that cannot be changed.

Flexible circuit: A vast array of conductors bonded to a thin dielectric film.

Hertz (Hz): The unit of measure for frequency named for Heinrich Rudolf Hertz, who was the first to conclusively prove the existence of electromagnetic waves.

LED: A light-emitting diode; a special type of diode that lights up when electricity passes through it.

Maintained switches: Switches that retain their state until they are actuated into a new one.

Momentary switches: Switches that stay in a particular state only as long as they are activated and return to their original state when released.

Ohm (Ω): The international unit of electrical resistance.

Photoresistor: A variable resistor controlled by light; also called a light-dependent resistor (shortened to LDR) a CdS cell (if it is made from Cadmium Sulfate), or a photocell.

Potentiometer: A variable resistor with two or three terminals and a sliding or otherwise movable contact element, called a wiper, which forms an adjustable voltage.

Resistor: A component that adds resistance to a circuit and reduces the flow of electrical current.

SMD: Surface-mount device; any electronic device that is made through a method in which the components are mounted or placed directly onto the surface of printed circuit boards.

Switch: A component that creates a physical break in a circuit, which prevents the flow of electricity.

Variable resistor: A component that changes its values for reducing the flow of electrical current depending on outside influence.

Voltage: The difference in electrical energy between two points, rated in Volts (V).

Watt: The electrical unit of power, named after James Watt, a Scottish inventor who made improvements to the steam engine during the late 1700s and helped jumpstart the Industrial Revolution.

3

How to Design with Conductive and Reactive Materials

This chapter focuses on conductive and reactive materials and how they can be used in fashion and accessory design. **Conductive materials** have the ability to conduct or transmit heat, electricity, or sound. Examples of these materials include conductive fabrics, threads, paints, and tapes. **Reactive materials** respond to UV light, temperature changes, and contact with water. These include thermochromic, photochromic, and hydrochromic inks, beads, and embroidery threads or yarns.

After reading this chapter you will:

- Understand which conductive and reactive materials have an application in the fashion and technology context and be familiar with their particular properties.

- Know the conditions under which thermochromic and hydrochromic inks transform and how to use them for textile application.

- Have the knowledge to use reactive materials in your garment or accessory designs.

Learning about the various conductive and reactive materials will inform your design process and help you create soft circuits for interactive garments with added electronics. Your collection is driven by your main concept; you must evaluate how these materials enhance it. Learning about various kinds of reactive materials and how they can be integrated into your designs will be key in creating a visually exciting and functional collection. The following questions should lead your work as a designer:

- Which conductive materials would help me create a closed circuit in which electricity will flow uninterrupted?
- Are there any traditional fashion design materials or hardware with conductive properties that can be integrated into a soft circuit?
- What materials will work to enhance my design ideas conceptually?
- How can I work with existing conductive materials to create enticing surface texture while maintaining a closed circuit?
- Can I augment the design to include conductive materials or hardware for switches?
- Which conductive or reactive components will work best with the chosen materials in my collection?
- Is the interaction of the reactive materials with the outside trigger relevant to the overall design concept of my collection?

Conductive Materials

Conductive materials vary greatly from a very thin sewing thread to woven fabrics, gels, paints, and tape, but they are all made from materials that can carry an electric current. Each material has different resistance and should be carefully chosen for optimal performance. The following materials are most commonly used in fashion-related projects.

Conductive Thread and Yarn

Conductive thread carries a current of charged electrons just like a wire does, and it can be spun from plated silver or stainless steel. The plated silver tarnishes easily, so it is best to use thread made from stainless steel. The type most often used in computational fashion projects comes in two- or three-ply and can be sewn by hand or used in a sewing machine. The thread comes in various sizes of spools or is already wound on a sewing machine bobbin (Figure 3.1). It is a bit stiff, not very strong, and tends to get knotted easily, but it works very

Figure 3.1 Two- and three-ply conductive sewing thread comes in large spools or wound on a bobbin.

well in a fashion garment. Because it is made of stainless steel fibers, it does not oxidize like silver does and it is safe to wash. Originally it was used for industrial applications, but as the fashion and technology market has grown, stainless steel thread can be easily found online through a variety of distributors.

In fashion and technology projects, conductive yarns are usually used for knitting, crocheting, weaving, and embroidery. The conductive yarn can be mixed with regular yarn and knit together to achieve different degrees of electrical resistance. Conductive yarns can be knitted into the tips of gloves to be used for touch-enabled devices, or they can be embroidered as a surface treatment and included in a concealed soft circuit. They can also be woven into a textile to create an embedded electronic circuit. A good example of a conductive yarn application is the International Fashion Machine's POM POM™ dimmer (Figure 3.2). The POM POM™ is made from conductive yarns and can be positioned in the same place a traditional switch would be, creating a textural accent in any color. Since it is composed of yarn, it is designed to be soft and pleasant to the touch. A user can turn on, turn off, or dim the lights in a room by gently pushing the pompom.

Figure 3.2 The POM POM™ Wall Dimmer was an early IFM product; it is a capacitive e-textile sensor made from conductive yarn that senses your touch. You can simply pat the conductive pompom to turn the lights on and off or hold on to it to set various dimming levels.

Figure 3.3 Conductive wool.

Conductive Wool

Conductive wool is made from very fine steel conductive filaments mixed with natural wool or polyester fibers (Figure 3.3). Wool works better for felting and can be used with great success for making pressure-sensitive variable resistors in soft circuits.

Conductive Fabric

Conductive fabrics can be woven or knit and vary in structure from very fine, see-through mesh to densely woven fabrics such as nickel/copper ripstop or even pure copper polyester taffeta (Figure 3.4). In most cases, they are washable and can be cut and sewn easily just like any other fabric. 100% surgical steel fibers can be woven into a strong but soft knit fabric that permits breathability, resists corrosion, conducts electricity, and is washable.

Some fabrics are blended or bonded with a natural material—for example, 100% pure cotton or conductive silver—to provide a soft touch and a pleasant feel to the skin. The bonded version will create a more functional blend. Many conductive fabrics can be laser cut in order to create custom or decorative electronic circuits.

Conductive Hook and Loop Fastener

A hook and loop fastener, similar to Velcro™, can be made into a **conductive loop fastener** when it is silver coated and thus made to conduct electricity (Figure 3.5). It is best used for making switches or connections that need to be opened and closed. You can also create your own conductive loop fasteners by weaving conductive thread in between the hooks and stitching some onto the loop side of a piece of Velcro™.

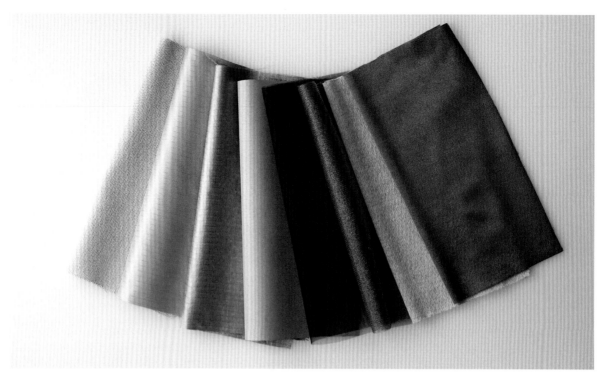

Figure 3.4 Conductive fabrics range from stretch knit jerseys to various woven fabrics, such as ripstop and silver plated nylon.

Figure 3.5 Conductive hook and loop fastener.

Conductive Tape

The most common **conductive tapes** are copper and aluminum. They are available in various widths, starting at ⅛" for the copper foil tape (Figure 3.6). Some have conductive adhesive and make a strong electrical connection simply by overlapping one end to the other. The tapes can be used to connect to a power supply and varied electronic components. Being adhesive, they are very easy to work with and can be attached onto different surfaces and at low temperatures can be soldered directly. Conductive tapes are great for prototyping as samples can be constructed quickly.

Conductive Paints and Inks

Conductive paints and inks are compounds that contain copper, carbon, or silver. They can be found in a liquid form, ready to be used with a brush, or prefilled in a pen (Figure 3.7). Various companies like Bare Conductive and CircuitWorks® sell such pens. The original purpose for CircuitWorks® pens was to repair electrical circuits, and they are not very reliable for fabrics. The Bare Conductive Bare Pens are a great electronics prototyping tool and work best on paper, cardboard vellum, wood, and some rubbers and plastics. The paint dries quickly at room temperature and can easily be removed with soap

Figure 3.6 Copper conductive tapes are available in various widths.

Conductive Materials   91

Figure 3.7 Bare Conductive paint is available in a jar to be used with brushes or for silk-screening; it also comes prefilled in a pen.

and water. For textiles it is recommended to test each fabric or material individually in order to achieve the best results. The advantage of Bare Pen is that the paint is nontoxic and water-soluble, so it can be used without gloves or mask. One known issue with such paints is that they tend to crack with wear and can interrupt the circuit. To avoid that, the pigments can be mixed with flexible materials such as fabric mediums, silicone, or latex. The newly created mixture has better flexibility when printed or painted on fabrics.

Metal Fasteners and Closures

In the fashion industry, traditional garment and accessory fasteners and hardware closures are metal and can easily be adapted for the making of electrical connections. Those can include zippers, with or without conductive tape; hooks and loops; hook clasps; and buttons.

Hook and Eye

The hook and eye is a traditional closure used throughout the fashion industry for fine garments (Figure 3.8). It can be found as single closure at the end of a zipper to secure it or in multiples of two or three at the back of a shirt or the waistband of pants or skirts. Hooks and eyes also come attached on two separate strips—one for the eyes and one for the hooks—and can be used in corsets or other delicate undergarments. The eyes can be straight or round: The straight eyes are used when the closure should be completely hidden, and the round eyes are used when the closure is slightly visible from the right side of the garment.

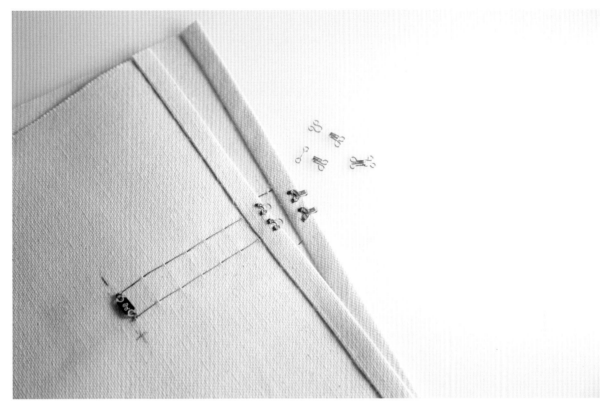

Figure 3.8 Metal hook and eye closures have conductive properties.

Hooks and eyes are made from nickel-plated steel, which makes them conductive; they can become a wonderful switch for a fashion garment. Do not use plastic-covered or painted hooks and eyes. The extra layer on top of the nickel will prevent the flow of electricity, and they will not be conductive. Do not forget to sew the hooks and eyes with conductive thread, or electricity will not flow through the soft circuit. See the tutorial at the end of this chapter for details.

Sew-On Snaps

Snaps are also widely used throughout the fashion industry for various garments, from light shirts to heavy outerwear (Figure 3.9). They vary in size from quarter of an inch to almost an inch and can be custom made. Use the nickel plated brass or brass snaps and avoid enamel coated and painted snaps, which will not be conductive. Make sure you sew the snaps with conductive thread. If you use a regular thread, you will create a break in the soft circuit and electricity will not flow through it.

Zippers

As reviewed in Chapter 2, zippers with metal teeth are excellent conductors of electricity and can be used for switches. Take a look at the zipper switch tutorials in Chapter 2 to see a few different applications and suggestions of use for them. Zippers with plastic teeth can also be used for momentary switches as long as they have a metal head. Please see the plastic zipper switch tutorial for details on how you can create one yourself.

Figure 3.9 Sew-on snaps with conductive properties.

Reactive Materials

Reactive materials are all materials that respond to an external trigger and transform in some way. They vary greatly, from textiles to threads, yarns, paints, resins, etcetera, and the materials discussed in this chapter can be successfully used for fashion or accessory design projects.

Thermochromic Inks

Thermochromic inks are pigments or dyes that change color as a result of a temperature change (Figure 3.10). The transformation happens at a predetermined temperature and can vary as defined by the manufacture of the inks. These pigments can change to a different color or to colorless. This property can be used to reveal another layer of ink underneath or the base textile. With this in mind you can keep layering inks with different temperature thresholds and create a dynamic visual print on a textile or a hard surface. Each layer would be activated at a different temperature creating an ever-changing design of your own.

Various types of thermochromic pigments can be used for painting directly on fabric or for textile screen-printing. In most cases they need to be mixed with a painting medium or a liquid binding agent if the pigment is in a powder form. Regular textile pigments can be mixed with thermochromic inks so the ink changes from one color to the other as it reacts to the temperature changes.

Wind Reactive Inks

Wind reactive inks are pigments developed by the British exploration house THEUNSEEN (see Figure 3.11). It features a form of wind reactive ink that changes color upon contact with the air around it. For this garment the material has been treated with liquid crystal to make it friction responsive. As a result varied shades of color appear and disappear as the wearer walks or the air moves around the garment. A gust of wind moving through the sculptured pieces will create visible color changes, indicating the invisible trigger.

Photochromic Materials

Photochromic materials change color with changes in light intensity or brightness. The normal state of the material is usually colorless or translucent, and it changes with exposure to a particular light source. When sunlight or ultraviolet radiation is applied, the molecular structure of the material changes and it exhibits a predetermined color. These changes are reversible; when the light source is removed, the color disappears and the material reverts to its original state.

UV Reactive Paints and Inks
UV reactive paint and inks respond to exposure to UV/sunlight. They can appear clear indoors and change to a bright color when subjected to light. They can also be blended to create new colors or change from one color to another (Figure 3.12). These pigments are very flexible in application and can be airbrushed, sprayed, painted, or screen-printed on fabric or leather. Color changing inks can also be found as oil based inks or heat transfer inks.

UV Color Changing Thread
UV color changing threads are developed to respond to UV light and transform from no color to vibrant bright colors or from one color to another. They range from embroidery to crochet thicknesses. Such thread and yarns are some of the easiest ways to incorporate this technology, allowing designers to embroider or knit elements into a garment or an accessory.

Plastic Color Changing Resin Concentrates
These resins can be found in clear-to-color or color-to-color change combinations and can be soft or rigid. Such resins can be incorporated in plastic products by mixing them in the process of plastic injection, blow molding, extrusion, etc. The end result can be any three dimensional product, including beads, jewelry, and shoe components like heels (Figure 3.13). Typically the color changes instantly when exposed to UV light and stays vibrant as long as it is exposed to light. Indoors the color transforms very quickly back to its original form. Such compounds are manufactured to last through thousands of color changes and remain consistent through their color modifications.

Hydrochromic Inks

Hydrochromic ink—as the name suggests—reacts as a result of exposure to water. Typically this ink is white when applied to the surface and becomes translucent upon contact with water. As soon as the surface dries, the ink reverts to its original state: white. The most common use for this ink is applying it on top of a multicolor print that is revealed as the ink becomes clear. It can be screen-printed or sprayed directly over the print.

The process of revelation can be permanent if the ink is irreversible. That means that the process is a one-time event that cannot be repeated. These inks can be successfully applied to both fabric and leather.

Figure 3.10 a, b These flowers are painted with thermochromic inks on muslin. They change from yellow to green and red. The green and red inks on this flower pattern are activated at 75 degrees Fahrenheit and become visible at that temperature.

Figure 3.11 a, b A garment from the Air Collection by the British exploration house THEUNSEEN. Its ink changes color upon contact with moving air.

Figure 3.12 Petal dress from Rainbow Winters. The dress is screen-printed with UV reactive ink. Indoors it remains orange; outdoors the sunlight activates the purple ink and the pattern changes to include both colors.

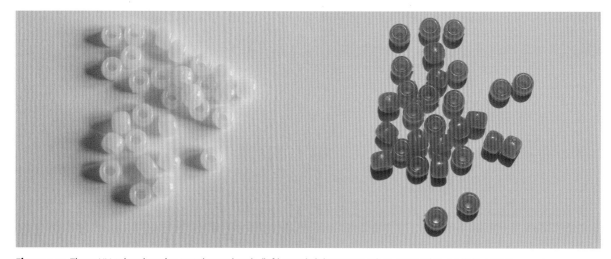

Figure 3.13 These UV color changing translucent beads (left) turn bright orange when exposed to sunlight (right).

Electroluminescence

Electroluminescence is the non-thermal generation of light from a material when a high electrical field is applied to it.[1] Commercially available electroluminescent materials are typically housed in plastic clear coating and come as either panels or wire. "EL" (for electroluminescent) products are typically manufactured to be powered by 110–120 Volts AC. Power inverters are available that take DC power via batteries and convert the power signal to AC. This allows for a compact power source that can be incorporated into a garment or accessory.

Because typical EL panels and wire are hard with plastic coating, they work best when incorporated into rigid structures. EL has been incorporated into clothing, mostly for clubbing and performance. EL display technology has been used in architectural and lighting design as well. Electroluminescent panels and wire can be purchased as a kit from companies such as Adafruit Industries and integrated into your project (Figure 3.14).

One particularly interesting application of electroluminescence is the Puddlejumper coat by

1. D. R. Vij, *The Handbook of Electroluminescent Materials* (Bristol, UK: IoP, Institute of Physics, 2004).

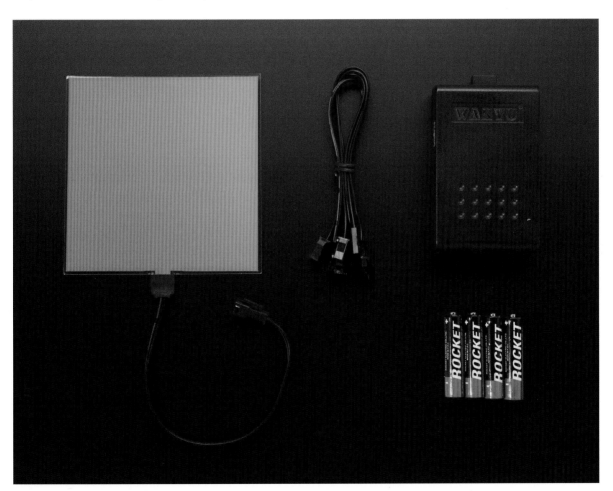

Figure 3.14 This Blue Electroluminescent (EL) Panel Starter Pack (10cm x 10cm) from Adafruit.com is ready to be integrated into a project.

Elise Co. (Figure 3.15). The Puddlejumper coat features hand-silkscreened electroluminescent panels that are activated when conductive ink sensors are triggered. The result is more subtle and textured than typical EL paneling. In an industrial context such imperfections would be viewed as a manufacturing flaw, but in this context the process of hand crafting EL components brings a unique aesthetic quality that enhances the project. A frequent critique of electronic materials is the lack of patina exhibited by the uniform plastic encasements of electronic components and the clean sterile aesthetic associated with technology. Elise Co.'s EL panels brings that unexpected element of the hand-made to computational fashion (Figure 3.16).

Figure 3.15 The Puddlejumper coat by Elise Co. The image to the left depicts the front of the coat with hand-silkscreened EL paneling. The back of the coat (pictured right) shows hand-silkscreened conductive inks arranged into sensors that detect the presence of water.

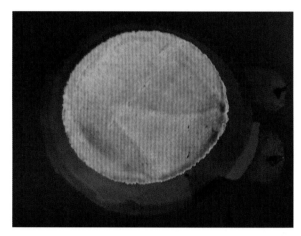

Figure 3.16 Detail of the hand-silkscreened electroluminescent panel on the Puddlejumper Coat. Commercially produced EL panels have a flat, uniform appearance. The texture of this panel would be considered a manufacturing flaw in an industrial context, but in the context of fashion it is valued as a design feature with subtlety and visual interest.

TUTORIAL 1:

SWITCH TUTORIAL—CONDUCTIVE HOOK AND EYE CLOSURES

This tutorial will guide you through the steps of adding two hook and eye closures to a simple soft circuit with one LED. This can also be referred to as a switch tutorial with hook and eye closures. You can modify this example to include more LEDs in parallel and, of course, change the shape of the circuit itself.

This tutorial gives you the basic skills to create your own conductive closures, but it should not restrict where and how you apply this technique. Keep in mind you should develop the circuit that you need depending on your design. For this example, we are working with two fabric swatches, but for your project you could be working on a closure for a sleeve or the back or front of a garment.

Whatever the case may be, make sure you carefully consider where the battery should be hidden, and try to have a short distance from the battery to the hooks. The shorter the circuit, the less chance there is of it being interrupted or breaking. We have outlined a rectangular shape for the conductive thread leads, but you can shape it in a circular manner and have the LED closer or further away from the closure. Whatever shape you choose, make sure that you create a complete and closed circuit.

Materials List

Two pieces of fabric, pencil, conductive thread, needle, two hook and eye closures, coin cell battery, battery holder, sewable LED. Yours may vary depending on your choice of fabric, as well as how many LEDs you use. For this example, we worked with one LED and created the circuit with hand-stitched conductive thread, but you can apply this technique directly on a garment or an accessory.

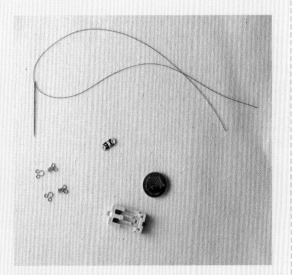

Figure 3.17 Materials.

Electroluminescence   103

Step 1 (Figure 3.18): A hook and eye is usually added to a clean finished edge of a fine garment. For this example, we folded inside and finished our edges with a blind stitch. You can either prep your fabrics in a similar way or, if you are using this technique on a garment/accessory, make sure you have finished edges at both ends of the connecting pieces.

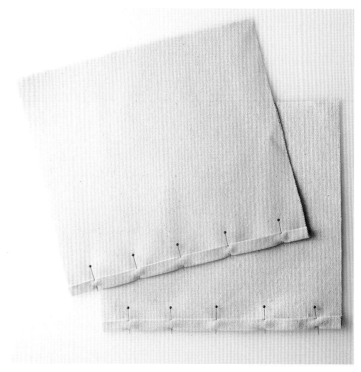

Figure 3.18 Step 1.

Step 2 (Figure 3.19): Carefully position your LED and mark where each loop/eye needs to be placed. Consider the whole circuit and make sure that the positive lead of the LED matches the positive lead of the battery case on the other side of the circuit. Then stitch the eye in place with the conductive thread with a blind stitch from the hook to the LED.

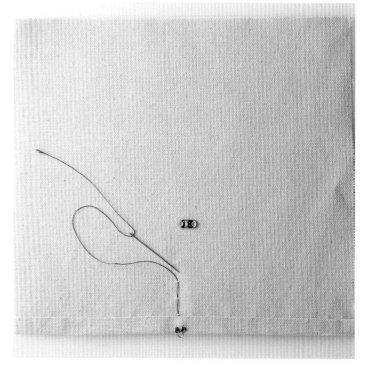

Figure 3.19 Step 2.

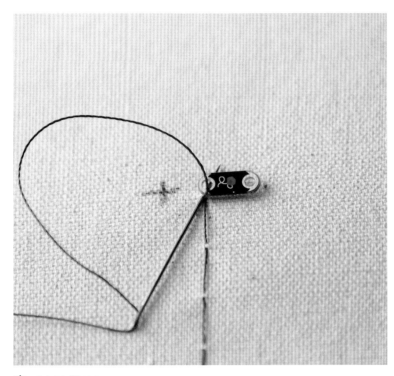

Figure 3.20 Step 3.

Step 3 (Figure 3.20): For this project we chose to stitch the LED on the inside so it would light up through the fabric. You can position yours on the inside or outside. Just maintain the positive lead of the LED connection to the positive lead of the battery. Since the markings are not visible once you turn over the LED, it is important to mark which side is positive and which one is negative. Here we marked with a pencil the "+" positive side of the LED to the left of the battery. Make sure you are not marking the right side of the fabric with ink you cannot remove later. To finish the stitching for this section, the conductive thread needs to be knotted at the LED lead.

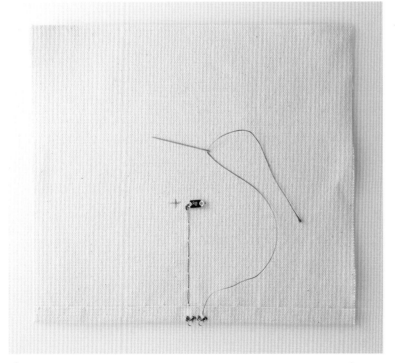

Figure 3.21 Step 4.

Step 4 (Figure 3.21): Stitch the connection from the LED's opposite end to the next eye, and knot the thread securely to each connection. Make sure you cut the loose thread short and you leave them on the inside, where they will not be visible.

Electroluminescence

Step 5 (Figure 3.22): The inside of your project should have clear markings for the LED positive and negative side and all threads should be knotted neatly. Make sure that there is at least ¼" space in between the two eyes. The eyes or the conductive threads from each one should not be touching or they will short the circuit.

Figure 3.22 Step 5.

Step 6 (Figure 3.23): For the opposite side of the closure we worked with two separate needles as it became easier to position both hooks at the same time and then continue to the LED. You may choose to attach one hook at a time and connect it to the battery case as we did in Steps 2, 3, and 4, or try this technique.

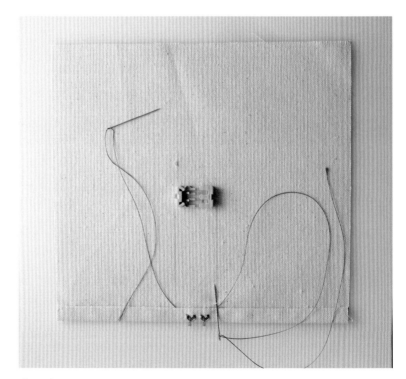

Figure 3.23 Step 6.

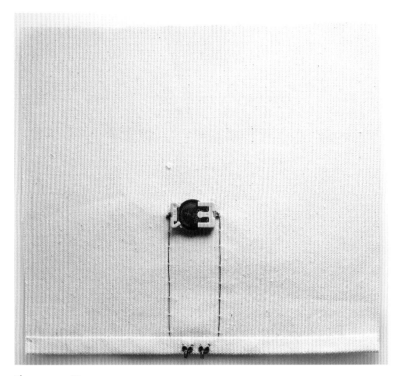

Figure 3.24 Step 7.

Step 7 (Figure 3.24): As you position and sew the battery case, make sure that the positive lead of the battery is oriented to match the positive lead of the LED and the negative with the negative.

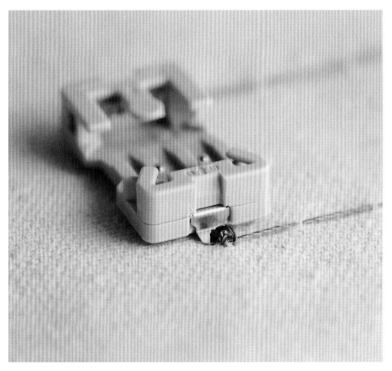

Figure 3.25 Step 8.

Step 8 (Figure 3.25): Loop each end of the conductive thread securely around the conductive leads of the battery. Each thread should be knotted on the inside of the garment or accessory. To ensure the knots do not get untied, you can add a drop of fray check or even nail polish to keep them from untying.

Step 9 (Figure 3.26): Place the two sides next to each other to check the positions of the hook and eye closures. They should match up perfectly. In this figure the hooks and eyes are not connected and the circuit does not light up. In other words, the switch is off.

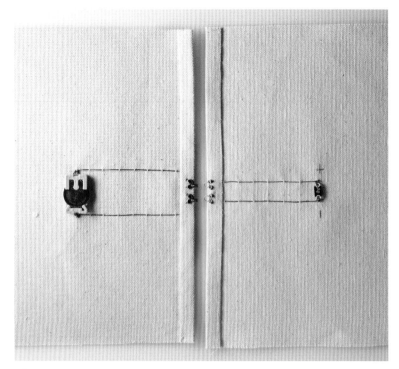

Figure 3.26 Step 9.

Step 10 (Figure 3.27): Connect the hooks and eyes to check whether the circuit closes. In this figure the hook and eye closures are hooked and the circuit is lit up. In other words, the switch is on. Note that both hook and eye closures need to be connected for the circuit to be closed.

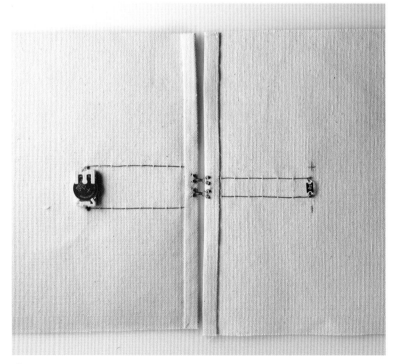

Figure 3.27 Step 10.

TUTORIAL 2:

SWITCH TUTORIAL—METAL SNAPS AS CLOSURES

This tutorial will guide you through the steps of adding two metal snaps to a simple soft circuit with one LED. This can also be referred to as a switch tutorial with snap closures. You can modify this example to include more LEDs in parallel and, of course, change the shape of the circuit itself.

Materials list

Two pieces of fabric, pencil, conductive thread, needle, two sets of metal snaps (not painted, rubberized, or enameled), coin cell battery, battery holder, and a sewable LED. Yours may vary, depending on your choice of fabric, drawing tools, and how many LEDs you use. For this example, we worked with one LED and created the circuit with hand-stitched conductive thread, but you can apply this technique directly on a garment or an accessory of your choice.

Figure 3.28 Materials.

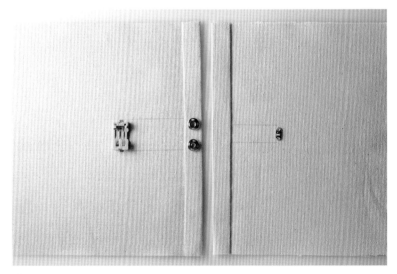

Figure 3.29 Step 1.

Step 1 (Figure 3.29): For this example, we folded inside and finished our edges with a blind stitch. Then we positioned all components exactly where we need them and marked with pencil where the conductive thread stitch lines will be. You can either prep your fabrics in a similar way or, if you are using this technique on a garment/accessory, make sure you have finished edges at both ends of the connecting pieces.

Keep in mind that the snaps on the LED side will be on the opposite side of the fabric and markings are not visible in this view.

Electroluminescence

Step 2 (Figure 3.30): Stitch the first snap in place and continue the stitch line with a blind stitch towards the battery case. To make the connection and keep the conductivity of the circuit, the conductive thread is needed only from one hole to the battery case. You could stitch the remainder of the holes of the snap with regular thread to match the fabric color.

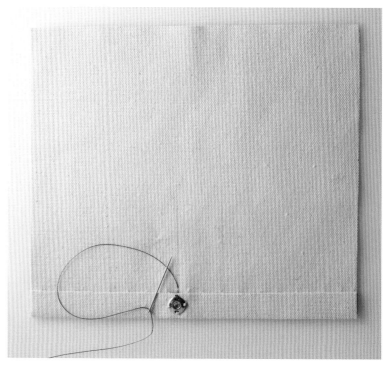

Figure 3.30 Step 2.

Step 3 (Figure 3.31): Secure the conductive thread at the end of the battery case lead with a knot, as shown in Figure 3.31. Add a dot of nail polish or fray check to secure the knot on the back.

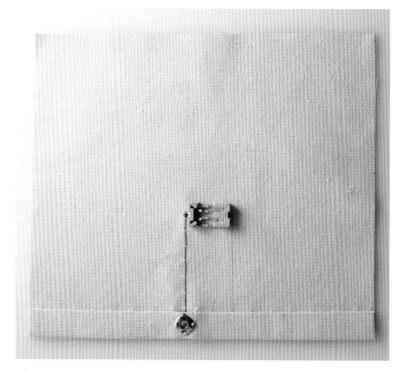

Figure 3.31 Step 3.

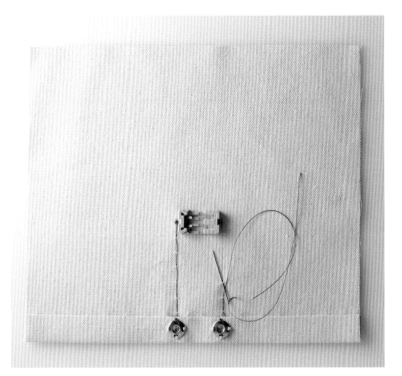

Figure 3.32 Step 4.

Step 4 (Figure 3.32): Carefully position the other snap and stitch it the same way along the marked line.

Figure 3.33 Step 5.

Step 5 (Figure 3.33): Flip over the finished piece to be sure it looks clean and polished on the right side of the fabric.

Step 6 (Figure 3.34): Position the male snaps on the right side of the fabric to match the distance of the female snaps. Mark each position.

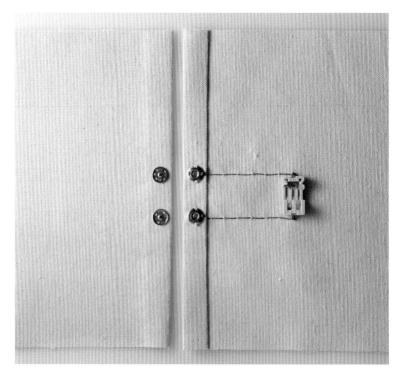

Figure 3.34 Step 6.

Step 7 (Figure 3.35): Blind stitch each snap with conductive thread from the hole closest to the circuit line. Note that you can stitch the remaining three holes with regular non-conductive thread to match the color of the garment.

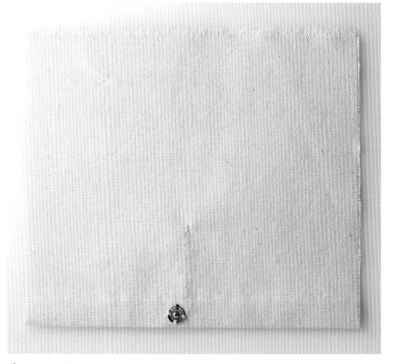

Figure 3.35 Step 7.

Figure 3.36 Step 8.

Step 8 (Figure 3.36): Flip the fabric on the other side to attach the LED. On this side the conductive thread stitching is visible, and the snaps are not seen here. Loop a few stitches around the conductive lead of the LED and finish with a secure knot and a touch of fray check or nail polish.

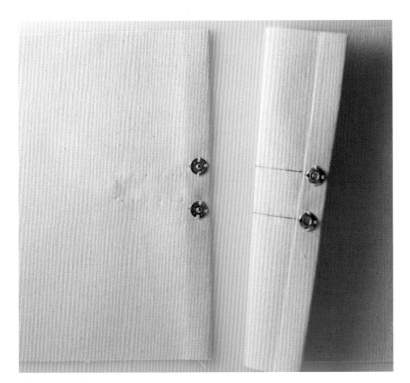

Figure 3.37 Step 9.

Step 9 (Figure 3.37): Repeat the same steps for the other snap and position both sides to check whether the snaps match up perfectly.

Step 10 (Figure 3.38): Connect the snaps to light up the circuit. In other words, turn the switch on. Note that both snaps need to be connected for the circuit to be closed.

Figure 3.38 Step 10.

Step 11 (Figure 3.39): Flip the swatch to view the closed circuit. In this Figure the snaps are closed and the circuit is lit up. In other words, the switch is on. Note that both snap closures need to be connected for the circuit to be closed.

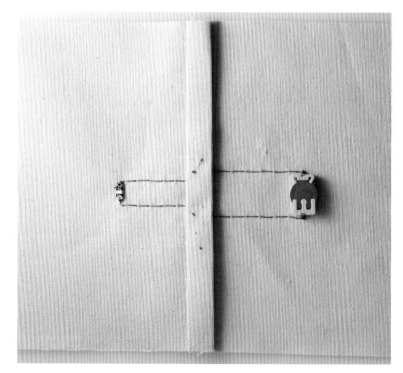

Figure 3.39 Step 11.

INTERVIEW
ISABEL LIZARDI (FROM BARE CONDUCTIVE)

Bare Conductive is a design and technology studio based in East London, UK. It started as a student project when Matt Johnson, Becky Pilditch, Bibi Nelson, and Isabel Lizardi developed Electric Paint as a way to soften electronics in a research project at the Royal College of Art in London. This non-toxic, electrically conductive paint is an innovative product that is used as a paintable wire. Once it is applied and it dries, it becomes conductive.

The company currently develops new products and kits and maintains a technology platform for people to get creative with technology and innovate. From paint to starter kits and hardware, Bare Conductive makes electronics fun and intuitive by blending design and technology. Electric Paint can be used on various fabrics or hard surfaces to create soft circuits with fashion applications, such as the flower-like circuit in Figure 3.40.

AG, KM: Please give us an overview of what Bare Conductive is.

IL: Bare Conductive Ltd. is a company focused on the development and manufacture of a new generation of electronics products.

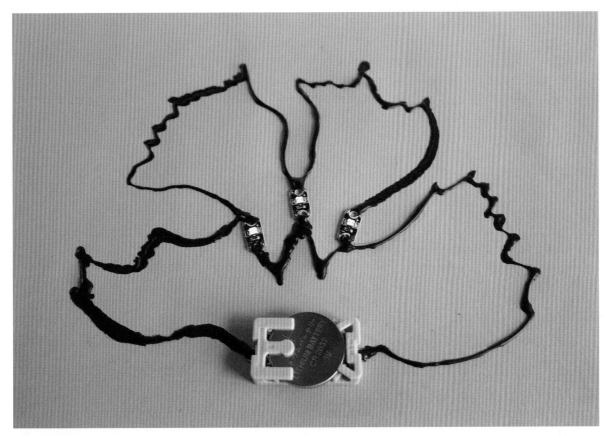

Figure 3.40 Soft circuit with LEDs connected in parallel, drawn with a Bare Conductive electric paint tube on paper. Created by Aneta Genova for this book.

The Company's purpose is to develop technologies that redefine the way that electronics look, feel, and behave, as well as how we interact with them. Bare Conductive does this by developing materials and hardware that allow electronics to integrate seamlessly into materials and surfaces that are not traditionally associated with electronics.

AG, KM: How was Bare Conductive started?

IL: Bare Conductive was founded in 2009 by four graduates from the Royal College of Art and Imperial College London. The company started out as a means to commercialize Electric Paint, a material developed as part of their Master's Thesis project. This material, a non-toxic, electrically conductive paint, was developed with the goal of creating a platform for the creation of graphical circuits and capacitive sensors on practically any substrate, from paper, to cardboard, fabric, concrete, and so on.

AG, KM: What other projects have you worked on in the past that lead to this product?

IL: All four founders come from distinctly different backgrounds and worked in a range of fields before Bare Conductive. I, Isabel Lizardi, originally studied Communication Design; however, before founding the Company I was working with Japanese crafts techniques and traditional manufacturing methods and materials. I believe this influence can be seen in the Company's focus on embedding and integrating technology into materials such as paper, wood, and fabric, which are not traditionally associated with electronics.

AG, KM: Why Electric Paint?

IL: Our focus has always been to create tools that are universally accessible, usable, and understandable. The reason for this is that we believe people of all backgrounds should be involved in the design and development of electronic products and interfaces. In recent history, this has been the realm of engineers and computer scientists; however, the products they have developed affect and reach a far more diverse group of people. Having come from a design background, we wanted to develop a tool that would be functionally and conceptually accessible to people of all backgrounds. Electric Paint is a material that both enables a wider audience to design and prototype with electronics, and opens up a new world of possibilities in terms of the materials and substrates that these can be built onto.

AG, KM: How has your product changed since the first inception?

IL: Electric Paint has been tweaked and refined over time to improve its performance; however, fundamentally it remains the same as when it was originally developed. Chemically it is a non-toxic, water-based paint that contains conductive particles and natural binders to keep them in suspension. Technically, it is an electrically conductive paint, which can be applied to most materials and used to create graphical circuits or capacitive sensors.

AG, KM: There are some really innovative uses of Electric Paint. What's the most exciting thing you've seen done with it?

IL: As our first product, Electric Paint has been on the market for several years now, so we've seen an incredible range of projects using it for very different applications. Two of the most memorable of these have been Patrick Stevenson-Keating's Liquidity, and Fabio Lattanzi Antinori's Dataflags: Lehmann Brothers. These two projects spring to mind because of how technically challenging and beautifully executed they were. The reason they are significant, though, is because they are perfect examples of how introducing a new material into the designer or engineer's toolbox—Electric Paint—can suddenly redefine the boundaries of how electronics can look or feel in our environment.

AG, KM: What other electronics can Electric Paint be used with?

IL: Electric Paint can be used alongside all conventional electronics tools and components. It simply allows for the extension of these onto surfaces that are not traditionally associated with electronics. It enables the expansion of our vocabulary of interactive surfaces onto materials such as fabric, paper, and even cardboard.

AG, KM: What applications for your product do you see in the fashion industry?

IL: I think our products offer the potential for the development of garments that are both intelligent and understated. They create a space for experimentation, not only in the function of smart garments, but also for the fabrication of wearable technology that is invisible and not stigmatizing to the user.

AG, KM: What advice do you have for fashion designers who would like to work with Electric Paint?

IL: My piece of advice is to experiment, iterate, and prototype. The best way to develop ideas using new materials is to test their limits and see what they can do. Concepts are only as good as what they turn into, so they must be made real and validated. When working with new technologies and materials, this is even more important, as there are more potential discoveries to be made.

For Review

1. What are conductive materials?
2. How do you define a reactive material?
3. What is conductive tape made of?
4. What kind of conductive fabrics can be found on the market?
5. How can you create your own hook and loop fastener?
6. What traditional fashion closures can and cannot be used as conductive closures? Why?
7. What external elements do thermochromic inks respond to?
8. What triggers a change in hydrochromic ink?
9. Can you layer color changing threads? Why or why not?

For Discussion

1. Which conductive materials are better used for prototyping and which ones should be used for the final garment or accessory and why?
2. Which conductive and reactive materials would you use together for a garment or accessory design and how would you use them?
3. Why would you use multiple layers of thermochromic inks for the same print or surface treatment? How would the blend affect the final result?
4. After reading this chapter will you approach your design process the traditional way—start with a concept for your collection and then look for the appropriate conductive materials to create the soft circuit—or are you inspired to tailor your designs to one or more of these newly discovered materials?

Books for Further Reading

Pakhchyan, Syuzi. *Fashioning Technology: A DIY Intro to Smart Crafting*. Sebastopol, CA: Make, 2008.

Vij, D. R. *The Handbook of Electroluminescent Materials*. Bristol, UK: IoP, Institute of Physics, 2004.

Online References

Berzowska, Joanna, XS Labs, http://www.xslabs.net, Web. 17 June 2015.

Mota, Catarina, and Boyle, Kirsty, Open Materials, www.openmaterials.org, Web. 17 June 2015.

Satomi, Mika, and Perner-Wilson, Hannah, KOBAKANT, http://www.kobakant.at/, Web. 17 June 2015.

Key Terms

Conductive fabrics: Woven or knit fabrics that vary in structure and carry an electrical curent.

Conductive loop fastener: A hook and loop fastener made from materials that allow it to conduct electricity.

Conductive materials: Materials with the ability to conduct or transmit heat, electricity, or sound; examples include conductive fabrics, threads, paints, and tapes.

Conductive paints: Compounds that contain copper, carbon, or silver; can be brushed or drawn onto fabric; and can conduct electricity.

Conductive tape: Usually made out of copper or aluminum, this tape can carry a current of electricity and act as a wire. It comes in various widths.

Conductive thread: Thread that carries a current of charged electrons just like a wire does; it can be spun from plated silver or stainless steel.

Conductive wool: Very fine steel conductive filaments, mixed with natural wool or polyester fibers, that can carry a current of electricity.

Hydrochromic inks: Inks that react as a result when exposed to water. They are usually transparent or invisible when dry and transform into a vibrant color when exposed to water.

Photochromic materials change color with changes in light intensity or brightness. The normal state of the material is usually colorless or translucent, and it changes with exposure to a particular light source.

Reactive materials: Materials that respond to UV light, temperature changes, contact with water, and other external triggers.

Thermochromic inks: Pigments or dyes that change color as a result of a temperature change.

UV color changing threads: Threads developed to respond to UV light and transform from no color to vibrant, bright colors or change from one color to another.

UV reactive paint and inks: Materials that change when exposed to UV/sunlight.

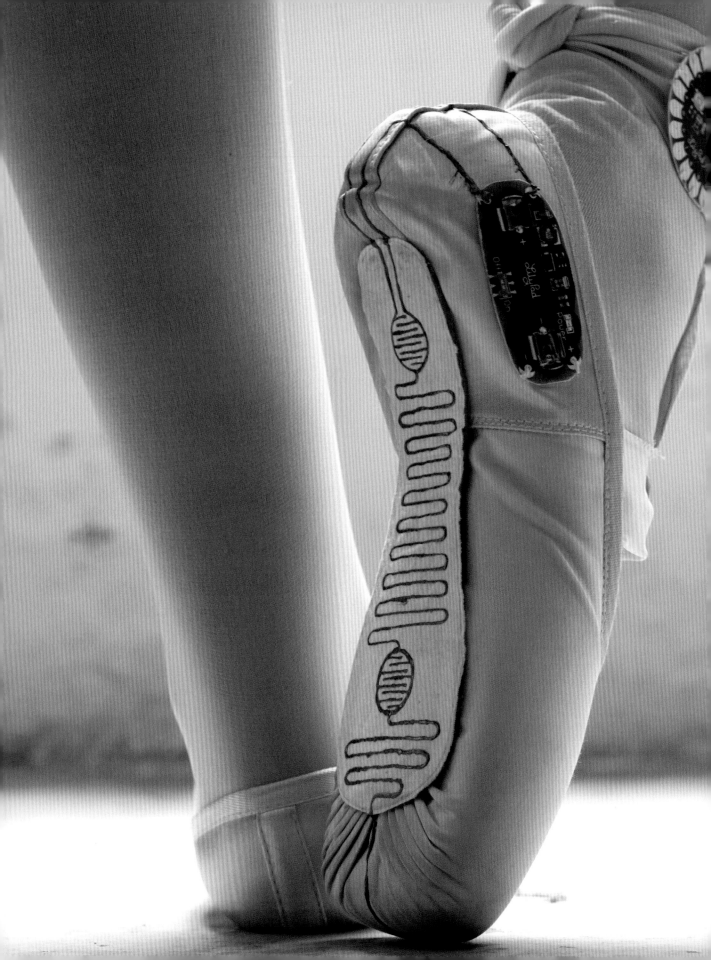

4

How to Design with Existing DIY Kits

This chapter focuses on understanding the history of the electronic Do-It-Yourself (DIY) space and its vibrant and active electronics community. It reveals the scope of sources and tools already created and readily available online and offline. It creates a knowledge base of the main established players in the field of DIY wearable technology and guides you through the various choices of videos, books, and tutorials, explaining how to work with existing kits and what their benefits are. This knowledge can be applied towards designing your own fashion apparel or accessories collections with electronics that you choose based on your well thought-out conceptual base.

After reading this chapter you will:

- Know the basic terminology for microcontrollers, electronic components, and peripherals.

- Understand the expanded tool kit you need for fashion and technology projects.

- Be able to conduct better informed searches in order to learn more about electronics and electronic kits.

- Know the community of makers and designers working in this area.

- Know where to buy the technology you need, how to work with it, and how to troubleshoot upcoming problems.

- Be able to further develop your projects through programming them.

- Understand the collaborative process between designers and technologists.

Most, if not all, of the design process for electronic and computationally driven fashion is collaborative because the complexity of knowledge and skills required to work in this area often exceeds the abilities of a single individual. This chapter will help you understand the language, tools, and materials needed so you can establish an effective collaborative relationship with technologists and programmers. With this knowledge you can build a collaborative team. Throughout this process you should be asking yourself some of the following questions:

- What electronics will fit best in my project?
- Which electronics can be handmade by me and which ones should be bought readymade?
- What additional tools do I need in order to successfully complete my project?
- Is there an **e-textile toolkit** that includes all the components I need and directions on how to implement them?
- Can I find solutions for the problems I am facing on online platforms and discussion forums?
- Does the branded website from which I buy the components offer tutorials I can use for my project?
- What new video tutorials have been posted on YouTube that can help me resolve my problems?
- What kind of collaborators do I need to complement my skills in order to successfully complete my design?

Overview of DIY Electronics

Recent years have seen a surge of interest in DIY electronics that are readily available and accessible to any designer with or without prior knowledge in the field. These developments in the DIY electronic space are invaluable for fashion and accessory designers who want to create projects with custom electronics. The process of creating computational fashion has become much easier to accomplish with the wider degree of accessibility of electronic toolkits. Numerous kits are readily available and can be bought through a variety of online stores, which also provide support and how-to tutorials for each specific kit.

In previous chapters we introduced basic electrical concepts, components, and various conductive and reactive materials. These materials are sufficient to create simple lighting effects and color changes, but to control or modulate input and output on a garment with additional circuitry and control we need more complicated systems.

Previously, working with electronics required specialized knowledge, but the advent of the DIY electronics community has broadened participation in the design process and thus led to development of lower-cost tools and kits. As a result fashion designers without engineering or technology backgrounds can now teach themselves how to build electronic circuits and program microcontrollers. The supportive communities of practice that provide workshops, feedback, and opportunities for collaboration already exist, and anyone can benefit from them.

For the student interested in electronic and computational fashion, the DIY electronics community provides a wealth of resources in the form of tutorials, materials research, and online forums. Many of these resources are available on the Internet, and there are several focused "how-to" books outlining how to "get started" with a particular technology or platform. In this chapter we will review some of the most active and important communities, which provide extensive support and ideas for a variety of fashion projects.

An Expanded Toolbox

As a designer, you can expect to use specific equipment on a regular basis. Your toolbox (Figure 4.1 a) includes not just actual tools but also a range of methods and approaches (along with the proper materials) that can be employed to realize a specific concept. For the practitioner who works with electronic and computational fashion, this toolbox is expanded to include tools from the DIY electronics community (Figure 4.1 b).

Electronic and computational fashion can be seen as a subset of what has become known as "physical computing." Broadly defined this means building interactive physical systems with hardware and software that can sense and respond to the analog world.[1] In the following section we will outline tools, resources, and development forms commonly used for physical computing.

---

1. Dan O'Sullivan and Tom Igoe. *Physical Computing: Sensing and Controlling the Physical World with Computers*. (Boston: Thomson, 2004).

Tools

Certain items are commonly accepted as part of the fashion designer's toolbox. Scissors, pencils, marking pens and chalk, measuring tape, sewing thread, and pins and needles are commonly associated with the profession. For the designer working with electronics and computation, there are new and unfamiliar tools, materials, and components that need to be included in an expanded toolbox.

Starter kits offered by vendors such as Adafruit and Sparkfun (and other similar retailers) combined with some select hand tools can be a simple way to acquire the basic materials needed. A typical starter kit will include electronic components (resistors, LEDs, pushbuttons, potentiometer), a prototyping board (breadboard), jumper wires, battery clip, USB cable, and a microcontroller development board (Arduino). Hand tools, such as wire strippers/cutters, needle nose pliers, a small screwdriver, and a multimeter, are also recommended. A multimeter (Figure 4.2) is especially useful when troubleshooting electrical connections and to check on the conductivity of materials you are working with. These starter kits are meant to give you the components needed for a variety of basic projects and offer most of the tools you need in the beginner exploration stage.

In addition to starter kits, many retailers in the DIY hardware space sell **fabricated kits** that can be customized or embedded in projects. These kits are much more narrow and focused in scope than starter kits and typically include only the products needed to complete a particular project. While limited in capability these kits can be useful for the electronics novice as they provide straightforward pathways to accomplishing a specific outcome (such as lighting LEDs or triggering sound) without having to configure and program a microcontroller board. If your design concept aligns with the featured project, then this can be a great way to acquire the needed materials.

a

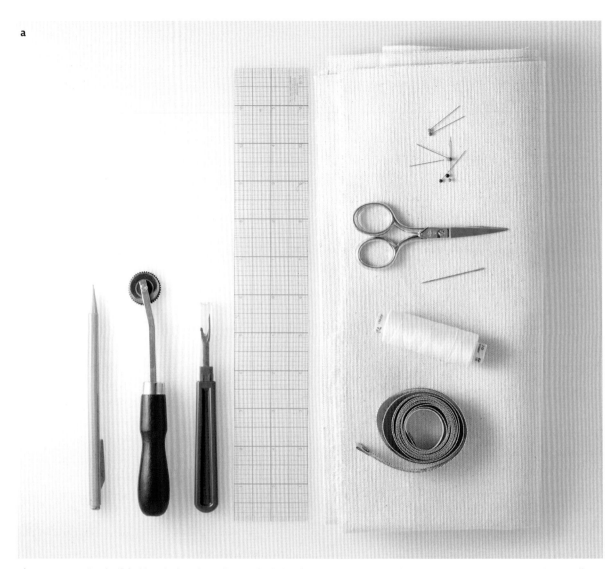

Figure 4.1 a A "typical" fashion designer's toolbox (a) includes, from left to right: pencil, tracing wheel, seam reaper, ruler, muslin for draping, pins, scissors, needle, sewing thread, tape measure, etc.

b

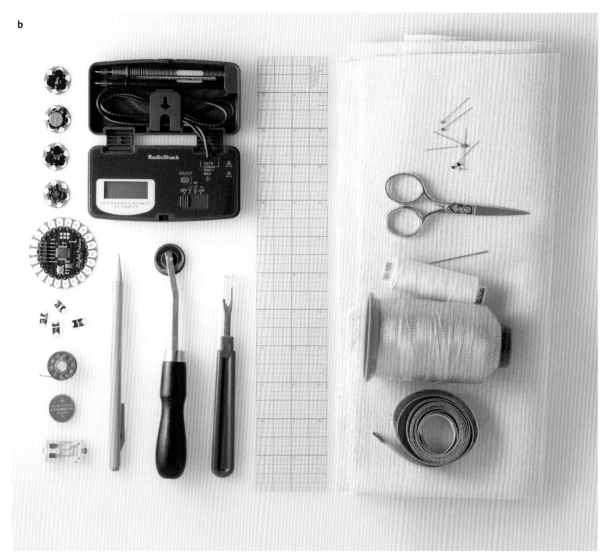

Figure 4.1 b An expanded kit should also include various microcontrollers and peripheral electronics, sewable LEDs, conductive thread, batteries, battery holders, and multimeter.

Figure 4.2 Multimeters come in a range of shapes and prices to suit a variety of budgets. The multimeter on the right folds into a portable case with all wires neatly tucked inside the box.

Many of these kits are offered through larger vendors, but can also be found by smaller kit developers who operate as small boutique retailers. One such kit (Figure 4.3) is offered through the invent-abling website. Deren Guler, a maker, tinkerer, designer, and physicist, is the creator behind invent-abling. She has worked on many educational and community-based projects and leads workshops at museums, universities, and other venues around the world.

Guler created invent-abling in attempt to fill the gap of low-tech toolkits for children, especially for young girls. Her website (invent-abling.com) offers a variety of kits which can be used for a wide array of fashion and technology designs. She has combined different materials, craft techniques, and computational methods to make accessible tools and compiled easy-to-understand directions for each set.

Such a kit can be very useful in your projects when it contains the components you need. You can use the provided electronics and test your idea on the supplied material, and then substitute the fabric with your own find for the final creation.

Figure 4.3 The invent-abling Smart Sewing kit includes, from left to right: conductive thread, metal snaps, sewable LEDs, battery holder, battery, and a piece of fabric. A direction guide can be downloaded from the invent-abling website.

Resources

The Internet provides access to a large number of tutorials, communities, and example projects that may be helpful as you develop your own electronic and computational fashion projects. Just like much of the content available on the Internet, these resources change over time. A website that is very active and current in one moment can be disabled in the next. This is to be expected in online as well as offline spaces, though the speed of turnover is accelerated on the Internet.

Nevertheless, there are countless and growing resources on the Internet for learning about DIY electronics as well as electronic and computational fashion, and you should be searching actively for the information you need. Though a specific URL may change, similar content will still be available elsewhere on the web if you know what keywords to search for. Search by keywords describing exact tools, components, electronics, and issues that you need to resolve to get the best results.

Once you find a trustworthy community, bookmark it and visit it often to get new ideas and learn how to troubleshoot common problems. Engage in conversations and post questions for your particular problems. Keep in mind that most problems have already been encountered by other makers, and there may already be a solution. Search through the existing posts first and then ask questions that remain unanswered.

DIY Electronics

For many years the Interactive Telecommunications Program at New York University's Tisch School of the Arts has maintained an online physical computing resource (https://itp.nyu.edu/physcomp/). This has been one of the more detailed and reliable websites for getting information about DIY electronics on the Internet. Anyone can acquire knowledge on how to get started with electronics as a beginner without a prior background or training. Other websites like Instructables (www.instructables.com) (Figure 4.4) and Make (www.makezine.com) outline how featured projects are constructed and built using DIY electronics.

Many of these projects are not fashion related, though electronic garments and accessories are periodically featured. Most of these projects come from makers who have no fashion design training and limited sewing experience, yet they can and often do provide novel concepts and methods for realizing a fashion design idea. These projects can act as resources for technology research allowing you to build your own knowledge of how to work with electronics.

Always start with a solid concept, then choose your materials and components based on that idea, and then work on how to incorporate electronics by way of searching through online sources or how-to books as described in this chapter. Do not expect

Figure 4.4 Instructables.com features a robust community of makers and a tremendous number of projects.

these communities to have a tutorial for the exact concept you are developing. Be flexible with your ideas to determine the best solution for embedding electronics, based on what works. Discard ideas that seem impossible to make as a beginner; implement tried and proven techniques until you gain experience and can solve more complicated problems.

You should also search through fashion and technology websites such as Fashioningtech (http://fashioningtech.com/), which compiles current projects from both fashion and interactive design. Various technology blogs such as Gizmodo (www.gizmodo.com) and Engadget (www.engadget.com) provide daily updated content related to technology and gadgets. Occasionally fashion and technology projects appear on these sites, and these may be sources for inspiration or ideas.

Lastly, developers of DIY electronics platforms will also provide their own "how-to" and "getting started" guides that typically focus on their product. The popular Arduino, an open-source physical computing platform based on a simple microcontroller board and a development environment for writing software for the board, maintains a website with forums and tutorials at www.arduino.cc. They have an extensive "learning" section with examples, references, and a playground section, where all users of Arduino contribute, and you can benefit from their collective research. The Arduino forum (http://forum.arduino.cc/) offers forum topics on installation and troubleshooting, project guidance, programming questions, general electronics, microcontrollers, and sensors, as well as e-textiles and craft and interactive art.

Likewise, The LilyPad Arduino (http://lilypadarduino.org/), a derivative of this board focused on e-textiles and electronic and computational fashion, maintains its own website focused on this specific board. Additionally, Sparkfun (www.sparkfun.com) and Adafruit Industries (www.adafruit.com), both distributors and manufacturers of DIY electronic components and kits, have extensive project tutorials that can be accessed.

One such project featured on the LilyPad Arduino website is the Electronic Traces project, conceptualized and designed by Lesia Trubat. (Figure 4.5) This design is based on capturing dance movements and transforming them into visual sensations through the use of the LilyPad Arduino technology added onto pointe shoes (Figure 4.6). As the shoes contact the ground while the ballerina is dancing, the LilyPad Arduino records the pressure and movement of the dancer's feet and sends a signal to an electronic device. The data can be seen as a graphic in a mobile application (Figure 4.7) and can be customized to suit each user through the various functions of the app. The final result is a beautiful visual representation of the movements that can be viewed by the users in a video format or as individual stills. Dancers can interpret their own movements and correct them or compare them with the movements of other dancers, as graphs created with motion.

Anatomy of an E-Textile Toolkit

Currently the LilyPad Arduino (Figure 4.8) and the Adafruit Flora are the two most well known microcontroller platforms developed specifically for the e-textile and electronic and computational fashion projects. Both are based on the design of the popular Arduino, a hobbyist microcontroller platform designed to make electronics design more accessible to the novice and beginner. These microcontroller platforms feature programmable devices that allow for the scripting of behaviors and responsive feedback, based on how they are assembled and programmed.

While specific brands and manufacturers of DIY electronics kits will undoubtedly change in

Figure 4.5 Lesia Trubat's project Electronic Traces features pointe shoes with the added LilyPad Arduino Technology.

Figure 4.6 Close-up of how the Arduino is added onto the ballet shoe.

the future, there are key features to these systems that are shared across all models. Both include a microcontroller and have the options to connect sensors, actuators, and peripherals, which can enable input and output and create preprogrammed garment behavior. While the LilyPad and the Flora are the two most commonly used microcontroller platforms available for e-textiles, in the future there may be other development platforms that designers will want to consider. Regardless of the type of platform available, e-textile toolkits will have some shared characteristics. Acquiring the knowledge of how to work with each one will help you control and program garment behaviors with various toolkits and software running on the microcontroller.

Syed Rizvi defines a **microcontroller** as a "small computer on a single integrated circuit (IC) containing a processor core, memory, and

Figure 4.7 Lucia Jarque's movements recorded and displayed on an iPhone with the help of Arduino technology.

Figure 4.8 The LilyPad Arduino was designed and developed by Leah Buechley and SparkFun Electronics. This is a microcontroller board specifically designed for wearables and e-textiles.

programmable input/output peripherals."[2] Development platforms for microcontrollers make access to the IC and its functions easier through providing physical connections to the board, supporting components, and software to make the board function. This means that individuals with less specialized knowledge can start working directly with prototyping and building their ideas.

Microcontroller platforms typically start with a main board (Figure 4.9) and have added components. In the case of the LilyPad Arduino or Adafruit Flora, the board has been specifically designed to aid in sewing it directly onto fabric, as evidenced by the through hole pads surrounding the board. Typically the power supplies are also provided. In the case of e-textile-oriented microcontroller platforms, various choices of power type are often provided, which can be selected not only for their electrical properties, but also for form factor.

The main board consists of the following components:

- Microcontroller—the brain
- Connector to power supply/battery
- USB port or FTDI connector port—connects to computer to program and/or monitor the device
- I/O Input and Output pins—enables the board to monitor sensors or controls actuators
- Analog pins—read analog sensors, and also have all the functionality of general purpose input/output (GPIO) pins
- Reset button

2. Syed R. Rizvi, *Microcontroller Programming: An Introduction*. (Boca Raton: CRC, 2012).

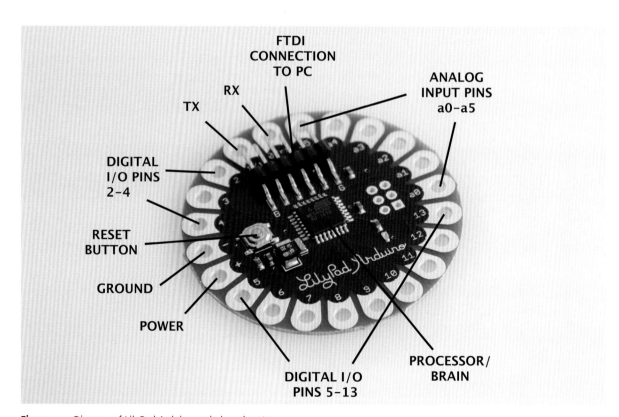

Figure 4.9 Diagram of LilyPad Arduino main board parts.

Lastly, input and output breakout boards are often included in a development platform. In the case of LilyPad and Flora, most of the sensors that are included as a branded part of the platform are simple, such as light or temperature sensors or simple switches. The output is primarily focused on sewable LED lights, although buzzers and vibration motors are also available. Keep in mind that other sensors and actuators can also be connected to these microcontroller platforms, but each sensor and actuator may have varying requirements for power and communication; thus, researching each specific component and its compatibility with the system you are using is essential.

Peripherals

The following LilyPad peripherals are specifically designed to be used for soft circuits and can be found in various websites and hardware stores. They are widely used in the DIY community and you can find a tremendous amount of information about them, including various projects implementing them and advice and support for each one. The LilyPad wearable e-textile technology is developed by Leah Buechley and cooperatively designed by Leah and SparkFun electronics. Some of those components include:

- Sewable LEDs (Figure 4.10)
- Sewable battery holders (Figure 4.11)
- Sewable sensors: temperature sensor, light sensor, accelerometer (Figure 4.12)
- Actuators: buzzer board (Figure 4.12)

Sewable LEDs

The LilyPad sewable LEDs are designed to have large connecting conductive pads on each side of the LED. They are easy to sew onto fabric, can be used in garments or accessories, and are even washable. They come in a range of colors and two sizes, as shown in Figure 4.10. The larger size is 5x11mm and the micro is a fraction of that. The micro LEDs are as close as it comes to sewing a small point of light onto your design with no extra weight.

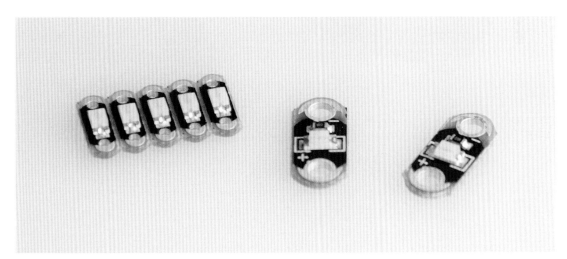

Figure 4.10 LilyPad sewable LEDs, in micro and regular sizes, come in a range of colors, from white to green, red, yellow, blue, purple, pink, and tricolor (not pictured here).

Battery Holders

When adding a battery to a soft circuit, you have a a few choices, as shown in Figure 4.11, but they all tend to be bulky and stiff as they wrap the battery in a plastic case. The advantage is that you can sew the holder directly on the fabric, connecting it to the LilyPad Arduino.

The most common battery holders are made for a 20mm coin cell battery. The LilyPad cases usually have four connection points (two positive and two negative) for sewing into your project. The LilyPad coin cell battery holder in Figure 4.11 (a) has a small slide switch installed directly on the board. It is positioned in line with the power so you can shut off your project and save batteries. The power supply pictured in Figure 4.11 (c) holds a single AAA battery. It is created to be as small as possible.

The gray plastic battery holder in Figure 4.11 (c) also holds a 20mm coin cell battery and allows the battery to be popped out with a single push of your finger. It also has conductive leads on each side so it can be sewn onto fabric. Beware: The holes on this holder are really small; you need a needle with a small head and you can only make a couple of loops with the conductive thread.

Figure 4.11 Sewable battery holders vary in size and shape, according to the batteries and the company that makes them. This image shows a power supply for a AAA battery (top) and two battery holders for coin cell batteries.

Sensors

Temperature sensors (Figure 4.12 top) detect touch-based or ambient temperature changes and output 0.5V at 0 degrees C, 0.75V at 25 degrees C, and 10mV per degree C. A sensor can be attached in a garment or an accessory as part of a soft circuit and programmed through an Arduino to trigger an output based on temperature changes.

An accelerometer (Figure 4.12 bottom) detects basic motion. This particular LilyPad accelerometer is a three axis one and can detect joint movement as well as inclination and vibration. Similar to the temperature sensor, it can be attached within a garment or an accessory as part of a soft circuit and programmed through an Arduino to trigger a desired output.

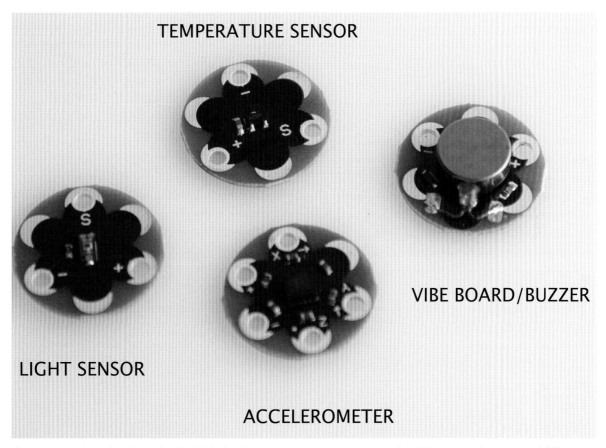

Figure 4.12 Various LilyPad peripherals. Clockwise, from top: temperature sensor, vibe board/buzzer, accelerometer, and light sensor.

Anatomy of an E-Textile Toolkit   135

A light sensor (Figure 4.12 far left) is very easy to use. If you expose the sensor to daylight, it will output 5V. If you cover the sensor with your hand, the sensor will output 0V. In a normal indoor lighting situation, the sensor will output from 1 to 2V. This can be programmed as a trigger or a switch for your soft circuit projects.

Other Peripherals

The LilyPad buzzer (Figure 4.12 far right) makes different noises based on the different frequency of I/O toggling and uses 2 I/O pins on the LilyPad main board. The noises it makes are not obtrusive but can be loud enough to hear inside a pocket. This buzzer cannot be washed.

LilyPad button board (not pictured here) is a simple button, designed to turn your project on and off. It is a momentary switch on a sewable pad with two conductive leads, very similar in its look to the sewable LEDs. The button closes when you push it and opens when you release.

Learning how to work with microcontrollers, sensors, and existing development platforms requires a steep learning curve and commitment. The greater the ambition of a project, the more knowledge is required. Because of this, many designers often choose to collaborate or assemble a team in which each member has specific knowledge to realize their ideas. That said, practical hands-on experience of assembling an electronic circuit and programming a microcontroller can assist the designer in better understanding the possibilities and limitations of working with these technologies and ultimately result in a conceptually stronger and aesthetically more nuanced piece.

Insight into the Collaborative Process between Designers and Technologists

Many fashion and accessory designers who are entering the world of wearable tech or computational fashion find that they need to collaborate with a technologist in order to successfully complete a project. Designers coming from traditional fashion education—even those with extensive experience in the industry—simply do not have the skills and knowledge to work with new technologies emerging every year. Developing electronics from scratch can be an almost impossible task.

DIY kits are a useful middle ground for prototyping concepts and interactions that would otherwise take enormous resources to pursue. That said, the skills required to do all of the work by oneself are still too vast. This is where collaborations and teamwork come in. If a project is relatively simple and limited in scope (for example, embedding LED lights into a garment, as outlined in previous chapters of this book), then the designer may learn and accomplish the tasks alone. However, for garments with complex sensing and response, an active collaboration is key for a successful completion.

Upcoming young designers who are currently enrolled in a fashion or accessory design program may want to take elective classes on physical computing or visit a local **Makerspace** to learn about working with electronics. Such endeavors will not make one a specialist in this area, but the hands-on knowledge acquired will make collaboration with another individual possible through shared experience and vocabulary.

CASE STUDY
CLIMATE DRESS BY DIFFUS STUDIO

The Climate Dress was created as a collaboration among Copenhagen-based design studio Diffus, Swiss embroidery company Forster Rohner, the Danish research-based limited company Alexandra Institute, and the Danish School of Design, and it included Tine M. Jensen as a fashion designer and Karin Eggert Hansen as a seamstress.

This dress represents an interaction between human and computer technology in a non-screen-based environment. The dress senses the CO_2 concentration in the air, and creates diverse light patterns according to the received information. Patterns can vary from light pulsations at a slow tempo to dynamic short and irregular ones.

Figure 4.13 The Climate Dress was created in a multidisciplinary collaboration among various experts.

Insight into the Collaborative Process between Designers and Technologists

As a partner, the Swiss company Forster Rohner integrated the soft circuits directly into the embroidery through an innovative production process. To better integrate the microcontrollers into the overall embroidery, Forster Rohner used a conductive thread quality that was very similar to a traditional embroidery yarn, which elevated the design aesthetic and allowed for a seamless integration from concept to execution. This way all functional elements were left exposed without impairing the design execution. See Figures 4.14 and 4.15.

Figure: 4.14 Close-up of the Climate Dress. Notice how the LilyPad Arduino becomes part of the surface treatment.

Figure 4.15 Detail of Climate Dress showing how the LilyPad Arduino was integrated on the surface as part of the embroidery.

A project like this one could not have been executed without the cultivated knowledge of diverse experts. That includes a trained fashion designer to conceive the dress; technologists with knowledge about microelectronics; an interaction designer; the e-textile innovation of a unique company like Forster Rohner, which has the existing manufacturing process to create a seamless integration of functional electronic components into a beautiful fabric; and the experience of a seamstress to put the whole garment together. The project was sponsored by Carlsberg's "Idé-legat" and Alexandra Institutes.

INTERVIEW
MICHEL GUGLIELMI (CO-FOUNDER OF DIFFUS DESIGN GROUP)

DIFFUS design is a Copenhagen-based design studio working with theoretical and practical approaches towards art, design, architecture, and new media using physical computing in an aesthetic context. Their mission is to support interaction between human and computer technology in a non-screen-based environment. In their work they combine traditional know-how and codified production processes with uncharted "soft" technologies.

Diffus was founded by Michel Guglielmi and Hanne-Louise Johannesen. Michel is an architect working with tangible media and interaction design. He has been teaching at the Royal Danish Academy of Fine Art, Schools of Architecture and Design. Hanne-Louise has a master's degree in Art History and has worked as assistant professor in Visual Culture at the University of Copenhagen. She now teaches at the IT-University, Denmark.

Figure 4.16 The Climate Dress is one of the award-winning projects created by Diffus design studio in collaboration with technologists and fashion designers.

AG, KM: How did the process of creating the Climate Dress start? Did you start with the interaction design concept or the style of the dress?

MG: We were invited to make a project during the COP15 Climate Summit in Copenhagen 2009. Very quickly we were considering the option of making a dress that had a relationship to the climate issues that were on the table for the actual event. We felt that it was appropriate to create a dress that interacts with its direct environment. The resolution came as a result of this decision. The decision concerning the aesthetic was motivated by several questions that we wanted to address:

- Can we imagine a garment combining the sensual qualities of traditional craft and the sophisticated properties of new technologies?
- Can we reveal the technologic complexity within the garment as an added aesthetic value? In doing so, the technologic features could become part of the overall design and acquire both a functional and decorative role.
- Can we propose a whole new relation to traditional embroideries and pattern making in which decoration and functionality get un-dissociated?

Here we think about Gustave Eiffel, and how he adopted the iron construction technology as structural and decorative element at the same time. In doing so Eiffel made it possible for an iron structure to become an aesthetic icon for modernity during the mechanical age.

But when you ask if we started with the interaction design concept or the style of the dress, the answer must be that you cannot fully start with one of the aspects—for us it is always a matter of combining different approaches to the actual project and making sure that they develop side by side. Designing with technology and interaction is, in our point of view, a complex integrated design process where complexity and openness have to be dealt with to the end. If not, there is a great risk that the technological part of the project will become slapped on top of an object and not integrated and designed with the object.

AG, KM: Logistically how did you organize the process? How often did all the participants meet and how did they share information?

MG: The time we had was very short, around two months. We had to take quick and bold decisions, which in our case favored the outcome of the project. Those sessions were intensive collaborations with a fashion designer and a group of students from the Danish School of Design. We also had regular meetings with two technicians from Alexandra Institute who were in charge of the actual implementation of the technology. These collaborations and the shared information among all the specialists allowed us to have in-depth knowledge on how to design with technology.

AG, KM: Was there a learning curve within the team when having to communicate and work with people from different disciplines (e.g., fashion designers with interaction designers?)

MG: At Diffus Design, we had experience working with both students (as teachers) and with technicians (as collaborators in former projects). Typically, Diffus acted as a mediator among the different members of the project team. The most interesting collaboration in this aspect was our industrial partner Forster Rohner. This was the first collaboration we had had with the fashion industry, so of course there was a learning curve in that aspect. But it has been exactly this learning curve and the results of it that have been the most fruitful, and it is still functioning as an asset in our further work with this specific partner as well as other industrial partners.

Also, the technicians from Alexandra Institute had a strong praxis that was based on regular collaborations with designers and creative people. The biggest challenge was probably experienced by the students who hadn't experienced a multidisciplinary approach of design. Naturally, students felt differently about such type of project. Most adapted quite fast but a few of them felt this way of working was too remote from their normal design practice.

AG, KM: Did the initial dress design change based on any feedback from the interaction designers or based on Forster Rohner's soft circuit layout requirements? Did the input of the fashion designer influence any part of the circuit layout?

MG: The interaction played a major role in the design of the embroidery and indirectly in the design of the garment on which the embroidery would be applied. More LEDs with more processing abilities could have been added but we needed to constrain ourselves to clear interaction rules between CO_2 level and LED patterns as pulse. Those clear rules influenced the design of the circuit layout as well as the design of the required algorithm. As mentioned earlier, all aspects of this project influenced each other, and therefore it was always important that none of the tasks were getting too far ahead.

Concerning requirements from Forster Rohner, we were quite surprised to discover that embroidered soft circuitry was a very versatile and context-adaptable solution for our design. Based on insights into the limitations of the embroidery process and the physical laws of electronics, we had to do very minor adjustments.

AG, KM: You used the LilyPad Arduino to realize this project. Could you tell us more about why you chose an existing electronics platform and what possibilities and limitations you encountered while working with an existing toolkit?

MG: We needed to work fast and had a short production time. LilyPad was perfect for this task. It was also interesting to explore the potential of its unique shape and pin layout.

Concerning the limitation: The performance level was adapted to the task; we had to rearrange the pin functionalities and we had to create extra LED-controllers because the LilyPad itself did not have a sufficient number of pins, but it was fairly easy to adapt it to our needs. Another possible limitation from the aesthetic of the LilyPad is that it bears specific connotations. Therefore, we have also worked on a customized solution for any copy we might make in the future. This version solves both issues of technical limitations and aesthetics.

AG, KM: What advice might you have for young fashion designers looking to get into this field?

MG: Partly the same advice as those for any good designer: Be open-minded and curious. Cultivate yourself and be ready to find inspiration in a wide range of different places and from our own past history. Use technology as an ingredient or material in the same way you use any other ingredient—and never as something added afterwards. Also, be prepared to work in a team in multidisciplinary settings. Finally: Be stubborn and don't give up.

TUTORIAL:
HOW TO USE A DIGITAL MULTIMETER CONTINUITY/ CONNECTIVITY TEST

One of the most important tools in your expanded toolbox is a multimeter. A multimeter can measure continuity, resistance, voltage, and sometimes even current, capacitance, temperature, etc. It really is a gadget with multitude of capabilities. There are a variety of multimeters, and they will all perform the basic function you need for your soft circuit projects. Most multimeters today are digital and are called DMMs (short for digital multimeter). Pick one that is most appealing to you in terms of size, functions, and ease of work. Make sure it has a continuity testing with a buzzer. Other functions you might want to have are as follows:

- Resistance test down to 10 ohm (or lower) and up to 1 Megaohm (or higher)
- DC voltage test down to 100mV (or lower) and up to 50V
- AC voltage test down to 1V and up to 400V (or 200V in the US/Canada/Japan)
- Diode testing

This tutorial will teach you how to check the continuity or the connectivity of the circuit. When the leads of components are electrically connected to each other and allow electricity to flow uninterrupted, it means that the connection is continuous. This process will help you determine whether and where a connection has been broken. A connection may break at a burnt LED or at a loose conductive thread that does not make a good connection with an LED or power source. To check such a flaw, you can use one of the basic functionalities of a multimeter, known as "continuity check."

You can check the continuity in a variety of materials and soft circuits. When you check for continuity/connectivity the multimeter sends a small amount of current through one of the DMM leads and to the circuit. If the current reaches the second lead and back to the multimeter with minimal or no resistance, then we have a connection. Sometimes a working connection is referred to as *short*. You will see this displayed on the multimeter screen in the following test when the connection is continuous.

Materials List

A multimeter, plain fabric, conductive fabric, and a soft circuit with LEDs you are already working on.

Step 1 (Figure 4.17): Switch on your multimeter and turn the DMM knob to **continuity mode**. (Some DMMs also double as diode testers and have **diode mode,** which is pictured here on both multimeters.) This mode is sometimes pictured with an audio icon.

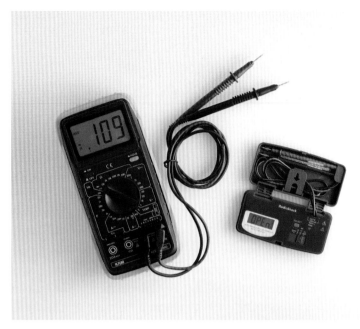

Figure 4.17 Step 1.

Step 2 (Figure 4.18): Start by testing a plain non-conductive fabric. Here we have a piece of muslin. Position it in front of you and hold each DMM lead in each hand.

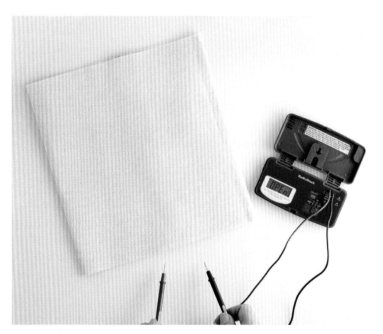

Figure 4.18 Step 2.

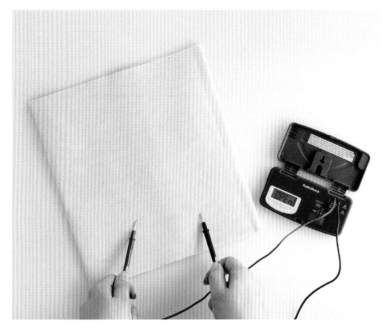

Step 3 (Figure 4.19): Touch the tips of your DMM leads to the fabric at the same time and hold them there. There is no circuit to be tested and the fabric is not conductive. The multimeter does not make any sounds and the screen displays the word *open* because there is no connectivity.

Figure 4.19 Step 3.

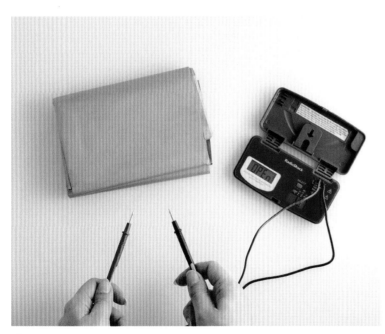

Step 4 (Figure 4.20): Here we have a piece of conductive woven fabric. Position it in front of you and hold each DMM lead in each hand.

Figure 4.20 Step 4.

Insight into the Collaborative Process between Designers and Technologists

Step 5 (Figure 4.21): Touch the tips of your DMM leads to the fabric at the same time and hold them there. There is no circuit to be tested but the fabric is conductive. The multimeter makes a sound and the screen displays the word *short*, signifying that the connection is continuous.

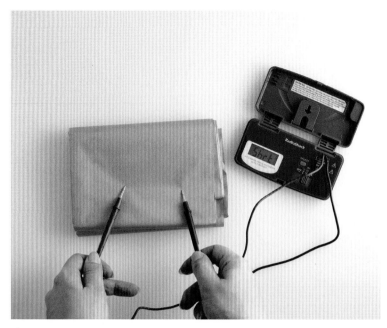

Figure 4.21 Step 5.

Step 6 (Figure 4.22): Test an existing soft circuit. For this test we are checking the connection of the conductive thread stitching. Touch the tips of your DMM leads to two different ends of the stitching line at the same time and hold them there. Make sure you are touching the conductive thread. The multimeter makes a sound and the screen displays the word *short* if the connection is continuous.

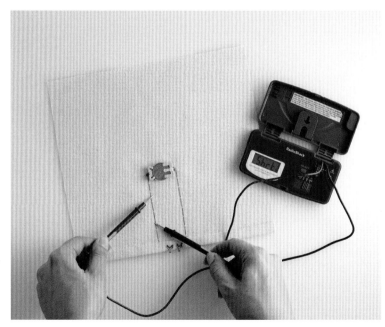

Figure 4.22 Step 6.

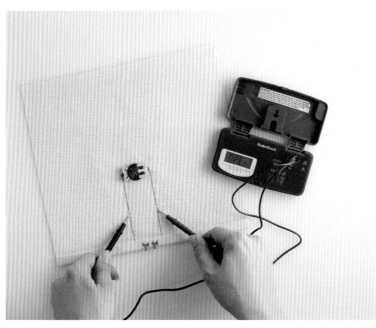

Figure 4.23 Step 7.

Step 7 (Figure 4.23): Test an existing soft circuit. For this test we are checking the continuity through the battery case, and the connections of the conductive thread to each lead of the battery case. Touch the tips of your DMM leads to two different ends of the stitching line at opposite ends of the battery case and hold them there. Make sure you are touching the conductive thread. In our case the multimeter does not make a sound and the screen displays the word *open*. That means that there is a break in the connection. It is possible that one of our knots is loose and not making a good connection. We can test each knot by touching the DMM leads to the thread stitching and the battery on each side of the battery case to determine which one is the faulty connection, and then repair it.

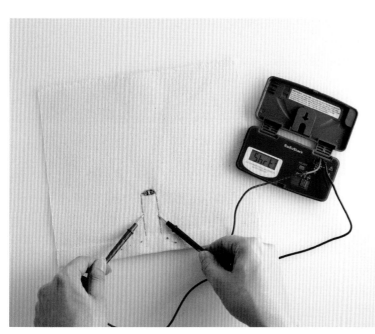

Figure 4.24 Step 8.

Step 8 (Figure 4.24): Test an existing soft circuit. For this test we are checking the continuity through the sewable LED, and the connections of the conductive thread to each lead of the LED. Touch the tips of your DMM leads to two different ends of the stitching line at opposite ends of the LED and hold them there. Make sure you are touching the conductive thread. In our case the multimeter makes a sound and the screen displays the word *short*. That means that the connection is continuous and electricity flows uninterrupted.

This technique allows you to check any circuit and troubleshoot all your soft circuits. It becomes extremely handy in complicated circuits with multiple LEDs or other components.

INTERVIEW

BECKY STERN (DIRECTOR OF WEARABLE ELECTRONICS AT ADAFRUIT INDUSTRIES)

Becky Stern is the Director of Wearable Electronics at Adafruit. She has been combining textiles with electronics since 2005 and helps develop the Adafruit FLORA wearable Arduino-compatible product line. She develops and publishes a new do-it-yourself craft+tech project tutorial and video every week and also hosts the YouTube Live show "Wearable Electronics with Becky Stern." Becky studied at Parsons School of Design and Arizona State University and currently teaches at the School of Visual Arts in the graduate Product Design program.

AG, KM: What is your background and how did you get involved in wearable technology and DIY electronics?

BS: I grew up around a lot of DIY; my parents were both avid home remodelers and from them I learned early on how to sew, knit, cook, take photos, and many more creative skills. They encouraged me to try whatever interested me, which was a lot of things. In high school I became very interested in video. I went to Parsons School of Design for my graduate degree and discovered electronics in a

Figure 4.25 Becky Stern, Director of Wearable Electronics at Adafruit Industries.

class called Physical Computing in the Design & Technology program. I made a set of plush illuminated steaks that became my first tutorial for MAKE magazine and cemented my niche of combining electronics with soft textile materials and traditional craft techniques.

AG, KM: You are currently the Director of Wearable Electronics at Adafruit. What does that job entail?

BS: I'm a media producer and my chosen subject matter is wearables. I create DIY projects for our customers to build using Adafruit hardware, then write, direct, shoot, and edit a video and photo tutorial on a rigorous schedule. Each week I host a live show on YouTube about wearables, as well as stock the Adafruit blog with posts about amazing DIY wearables and industry news. I also oversee the Adafruit YouTube channel and help out where I can with many other creative endeavors going on at Adafruit. I'm a Jane of all trades. I spend a lot of time in Final Cut Pro, on email, and looking through the viewfinder of my DSLR.

AG, KM: What are you currently working on?

BS: It changes every single day and every single week—I would say check the Adafruit blog and weekly wearables show!

AG, KM: You are very involved with the DIY electronic craft and wearable technology community. What do you think is the appeal of this area and why are people attracted to it?

BS: For me, I really enjoy experimenting with technology, and crafting is a fun and creative way to play and learn, partly because I came into electronics with a lot of craft skill background. So I think folks like me are attracted to wearables because it can introduce a textiles expert to electronics in a way that lets him or her experience almost immediate success in a field that can seem intimidating. For others I'm sure it's exciting to wear their electronics projects out in the world instead of leaving them at home on the workbench. Wearables help us express ourselves through fashion, which attracts creative makers who have an end-goal greater than just to learn Arduino.

AG, KM: What are your thoughts on the current tools available to create electronic fashion?

BS: It's the best time to be making DIY wearables. One of the reasons I joined Adafruit was to help create the FLORA line of boards and sensors to specifically make it easier than ever to put microcontrollers, lights, and sensors in garments and accessories. There is more choice than ever when it comes to tools and kits for e-textiles, and I'm so glad that we now have stainless steel thread instead of the old tarnish-prone silver stuff I used when I was getting started.

In addition to the physical products available, I think there is also a great wealth of information and tutorials online for DIY wearables right now. It's likely that if you have a project in mind, you can find a tutorial that will get you 80% of the way there, which was unprecedented in wearables until 2013.

AG, KM: How have these tools evolved to what is available today?

BS: That would require an entire book devoted to really getting an idea of how far we've come with the tools—in fact, a new book just came out about this! It's *Building Open Source Hardware: DIY Manufacturing for Hackers and Makers*, by Alicia Gibb.

AG, KM: What are the advantages and disadvantages of working with pre-existing systems like the LilyPad or the Adafruit Flora?

BS: The biggest advantage is you can get really far, really fast, using something that was already specifically designed with wearables in mind (and beginners!).

AG, KM: As part of your work at Adafruit you have developed an extensive set of tutorials for working on electronics and wearables. What is your favorite and why?

BS: Out of the hundreds of tutorials I've written, it is impossible to choose a favorite. However, my most popular wearables projects are the step-activated firewalker sneakers, GPS bike helmet, and volume-reactive necktie.

AG, KM: What technical innovations would you like to see in the area of soft circuits and electronics?

BS: I'd like to see more materials that bridge textile and technology, like Plug & Wear's textile perfboard that lets you solder components directly to it. I think it will take more innovators that bridge fields (like an electrical engineer who inherits the family's textile mill, for example) to bring new innovations to market.

AG, KM: What advice can you offer to young designers interested in this field?

BS: Don't be afraid to create your own path. Just follow what interests you and it will become your field of expertise. Fail fast by getting ideas prototyped as quickly/crudely as possible, then revise many times. Create swatches before embarking on a big or labor-intensive project to save yourself from heartbreak. Find collaborators with skills that complement your own.

AG, KM: What advice might you have for fashion designers who might be hesitant to work with technology and electronics?

BS: We have written dozens of tutorials for you to start with! Start small, and expect to fail on some early projects. Electronics and programming aren't any harder than putting in a zipper or other technique you already know, but like most worthwhile things, they take practice to master. Check out the weekly Adafruit electronics Show & Tell on Google+ Hangouts where we have seen tons of folks learn and excel using our tutorials. The community online is a great resource for getting support for your ideas and creations!

For Review

1. What is the difference between a typical fashion designer's toolbox and an expanded one for the needs of fashion and technology?
2. What are some of the most reliable online sources of information about DIY electronics?
3. What are components of a LilyPad Arduino board?
4. How can the battery holder act as a switch in a soft circuit?
5. What is a microcontroller peripheral?
6. What kinds of sensors are available from the LilyPad platform?
7. How do kits help the design and development process for fashion designers?
8. What is the purpose of a multimeter?

For Discussion

1. What kind of collaborators would you need for your projects?
2. Which sensors can you envision used in your projects and how?
3. Have you encountered a DIY project with embedded electronics? If so, what did you learn from it?

Books for Further Reading

Buechley, Leah, Kanjun Qiu, Jocelyn Goldfein, and Sonja De Boer. *Sew Electric: A Collection of DIY Projects That Combine Fabric, Electronics, and Programming*. HLT Press, 2013.

Eng, Diana. *Fashion Geek: Clothes Accessories Tech*. N.p.: North Light, 2009.

Gibb, Alicia. *Building Open Source Hardware: DIY Manufacturing for Hackers and Makers*. Upper Saddle River, NJ: Addison-Wesley, 2015.

Hartman, Kate. *Make: Wearable Electronics: Design, Prototype, and Wear Your Own Interactive Garments*. Sebastopol: Maker Media, 2014.

Lewis, Alison, and Lin Fang-Yu. *Switch Craft: Battery-powered Crafts to Make and Sew*. New York: Potter Craft, 2008.

O'Sullivan, Dan, and Tom Igoe. *Physical Computing: Sensing and Controlling the Physical World with Computers*. Boston: Thomson, 2004.

Pakhchyan, Syuzi. *Fashioning Technology: A DIY Intro to Smart Crafting*. Sebastopol, CA: Make, 2008.

Rizvi, Syed R. *Microcontroller Programming: An Introduction*. Boca Raton: CRC, 2012.

Stern, Becky, and Tyler Cooper. *Make: Getting Started with Adafruit FLORA: Making Wearables with an Arduino-compatible Electronics Platform*. Sebastapol: Maker Media, 2015.

Key Terms

Continuity mode: This mode of a multimeter tests the resistance between two points and tells us if the two points are connected electrically. If they are, then the multimeter emits a tone. This test helps ensure that connections are made correctly between two points.

Diode mode: A diode is an electrical component that only allows for electricity to flow in one direction. This test will show the proper direction in which to connect an LED. It is most often used in soft circuits to test if an LED is oriented correctly if there are no markings or if the + and − symbols cannot be seen.

E-textile toolkit: Typically a kit that contains all the components and fabrics to create a simple soft circuit in order to make the acquisition of these components easier.

Fabricated kits: Narrower in scope than starter kits, these typically include only enough products to complete a set activity.

Makerspace: A physical community space where people who consider themselves makers can get together and share resources, such as digital fabrication tools, electronics, and knowledge, to collaborate with each other.

Microcontroller: A small computer on a single integrated circuit that includes a processor core, memory, and programmable input/output peripherals (as defined by Syed R. Rizvi in *Microcontroller Programming*).

Starter kits: A typical starter kit will include electronic components (resistors, LEDs, pushbuttons, potentiometer), a prototyping board (breadboard), jumper wires, battery clip, USB cable, and a microcontroller development board (Arduino).

5

Introduction to Digital Fabrication

This chapter gives you an overview of digital fabrication and introduces concepts, vocabulary, and tools for designing using digital fabrication technologies. Technologies covered include 3D printing, laser cutting, and aspects of manufacturing on demand.

After reading this chapter you will:

- Have a better understanding of how digital fabrication presents an opportunity for fashion designers to generate form and material using new technology and how this process breaks away from established methods of construction and design.

- Be able to contemplate digital fabrication methods and design structures in your work.

- Understand how the impact of 3D printing is creating new demand for designers with an understanding of these technologies, and the ability to manipulate 3D models and forms, along with the disciplinary understanding learned through training in fashion design.

- Know the various fabrication tools and the requirements and affordances for rapid prototyping vs. large-scale production; be able to recognize which machine would fit a specific set of needs better and use available resources as needed.

- Learn what new skills need to be acquired in order to work with digital fabrication.

- Understand how collaborative process can and should be used throughout the design and prototyping process.

Designers coming from traditional fashion education and even with extensive experience in the industry may not have the skills and knowledge to keep apace of rapid development in digital fabrication. 3D printing, for example, requires the user to work with 3D modeling software to generate each prototype. This process has its own innovation cycle and requires specific machines with rapidly changing capabilities. Laser cutting requires each file to be created in a specific format for etching or cutting each line and demands knowledge of how various laser cutters and materials give different aesthetic outputs.

For upcoming young designers who are currently enrolled in a fashion or accessory design program, it is advisable to take various elective classes on the most current software in order to acquire the needed knowledge. Programs like Solid Works and Rhino are some of the most popular and widely used software for 3D modeling. They are traditionally used in product design and architecture, but their application in fashion design, and especially for jewelry, footwear, heels, or hardware prototyping, is invaluable.

This chapter provides an overview of digital fabrication technologies that you can and should apply towards designing your own fashion and accessories collections based on your conceptual theme.

Overview of Digital Fabrication

Digital fabrication is a term that commonly refers to a fabrication process in which a machine is controlled by a computer. The finishing touches can range for each individual product, but the unifying aspect is that the machines can be programmed by the designer to output consistent product.

In 1952 researchers at MIT wired an early computer to a milling machine,[1] creating the first digital fabrication tool. Since then the field of digital fabrication has expanded exponentially, with a wide range of techniques, machines, and processes becoming available. For the designer this creates an opportunity to work with new materials and methods for creating garments and accessories. Many of the tools and technologies now available have their roots in other design disciplines or come from the popularization of small-scale and personal manufacturing.

Digital fabrication can be used for both prototyping and finishing, and the choice of certain machines or services will often depend on where in the production cycle a designer is working. As with all of the practices outlined in this book, collaboration becomes a key

1. Neil Gershenfeld, "How to Make Almost Anything," (*Foreign Affairs* 9, no. 6, 2012: 43–57) accessed August 2, 2015, http://cba.mit.edu/docs/papers/12.09.FA.pdf

Figure 5.1 a, b A laser cut leather dress by threeASFOUR (a) and a 3D printed dress from the Spring 2014 show (b) (threeASFOUR runway show, Mercedes-Benz Fashion Week, September 08, 2013).

feature in successful design outcomes. In the case of digital fabrication, designers will often work with technologists or even professionals from other design disciplines, such as architecture or product design. It is already evident that fashion education, in addition to product and architecture, has begun to specialize in this area, adding momentum to newly forming job roles within the industry as digital fabrication techniques begin to proliferate widely.

In previous chapters we introduced basic electronics, reactive materials, and DIY toolkits. As a computer-controlled process, the advances in digital fabrication have been influenced by the rapid changes in technology. For example, the same momentum in the Maker movement that has propelled advances in DIY toolkits has also increased the accessibility and affordability of 3D printers. And yet, digital fabrication techniques have their own specific set of design challenges that do not necessarily include electronic components or computationally interactive functionality. While many designers opt to collaborate or contract out their digital fabrication needs, it is important for you to have working familiarity with digital fabrication technologies in order to best understand how to integrate them into the design process.

The software and machinery used to digitally fabricate is available at different degrees of refinement and price points, making hands-on learning in this area much more attainable. As with the DIY community for electronics, there are several online forums, workshops, and conferences that exist for the purposes of learning and support. Various companies now offer a range of services that can meet nearly any budget. Designers are increasingly turning to digital fabrication techniques in order to realize their collections.

Along with those needs, the specific affordances of digital fabrication have inspired new forms and new methods for design. Both the ability of a laser cutter to produce intricate cutouts on fabric or leather and the capability of a 3D printer to fabricate complex three-dimensional shapes that would normally be prohibitive to create by hand are creating new opportunities for designers. Each one of these processes allow individual designers to create highly customized materials and prototype components that otherwise would take more time and money and would involve a complex alignment of multiple partners. With the cost of both laser cutters and 3D printers falling, they are becoming more common, and universities as well as individuals are adopting the technology for wide use in their studios.

Types of Fabrication Machines

As mentioned above, digital fabrication is a term used to encompass a range of computer controlled fabrication techniques. It encompasses many kinds of machines and processes.

It is important to note that digital fabrication machines are divided into two main subtypes with different design affordances:

- **Subtractive fabrication** machines chisel or cut away at a block or sheet of material. Laser cutters and CNC routers fall into this category.
- **Additive fabrication** create objects through adding small amounts of a material until an object emerges. 3D printers are an example of an additive digital fabrication technology.

Each fabrication type requires some sort of software file specifying the type of object to be created. The type of file needed can be different for each type of machine. Some machines only accept a specific type of file; others may be able to translate multiple types of files into the instruction set necessary to generate the final object.

Of the various digital fabrication machines out there, laser cutters, CNC routers, and 3D printers are becoming more commonly used in fashion. We will discuss some of the design possibilities of each of these machines in the following sections. While there are other digital fabrication machines and tools that may be applied to fashion, we will focus on these three. It is important to note that while many digital fabrication tools were created with other applications in mind, they have been adopted in the product design and fashion design fields. In addition, while we focus on laser cutters, CNC routers, and 3D printers in this particular book, there may be other digital fabrication tools emerging that will be worth considering in the future.

Laser Cutting

Laser cutting is defined as a subtractive digital fabrication process. A laser cutter uses—as the name implies—a laser to melt, burn, or vaporize a sheet of material laid out on a flat bed. This can include plastic, metal, paper, cardboard, leather, or a variety of textiles. The pattern that the laser cutter will trace needs to be formatted in a digital file. The exact format of the file depends on the brand of the laser cutter you will use, but generally an Illustrator file indicates cuts and etching. Sometimes these are denoted through colored lines and/or stroke thickness.

Laser cutters allow for both through cuts and surface treatments (i.e., etching), though testing materials beforehand is key because the laser can create burn marks on some types of surfaces (for example, leather). Laser cutters are particularly beneficial for creating intricate detailed cutouts on acrylic (Figure 5.2 and Figure 5.3) and leather (Figure 5.4). Trying to create such details by hand would be either too complex or too time-consuming to be feasible, and the degree of complexity achieved by the machine is unmatched. Designers often use laser cutting to create custom clothing tags or labels, as in Figure 5.5. These are just some of the inventive and functional uses of the laser cutter for the fashion industry; others can include cutting out and adding branding elements on hangers and display materials (see Figure 5.6).

CNC Router

A **CNC router** is a computer controlled cutting machine. While originally used for wood and metal, the CNC router has been increasingly used in fashion for large-scale pattern cutting. It allows for multiple layers of fabric to be piled on each other and then cut with one pass of the machine. This is a time-saving tool that has great accuracy.

A CNC router can also be used to implement surface treatments. The CNC router is preferred over a laser cutter when burn marks from the laser might be evident after the laser cutting process (such as on a material like leather). Leather or fabrics with thicker weaves tend to work best on the CNC. A loose weave textile will tend to pull and move around and

Figure 5.2 Laser cutter in the process of cutting out designs on acrylic.

Figure 5.3 Three layers of laser cut acrylic shapes joined together with a grosgrain ribbon for a necklace designed by Aneta Genova.

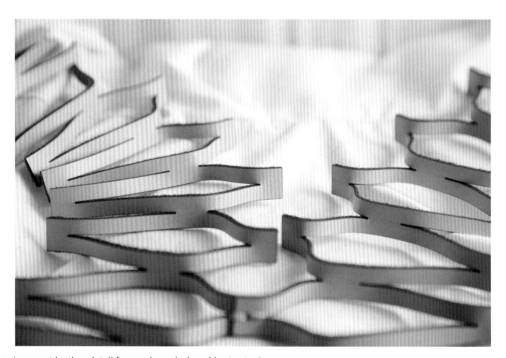

Figure 5.4 Laser cut leather detail from a dress designed by Aneta Genova.

Figure 5.5 Clothing label from Laura Siegel's clothing collection. This thick paper label is etched with a laser cutter. In this case, the "burnt" look works with her concept and enhances the aesthetics of her collection.

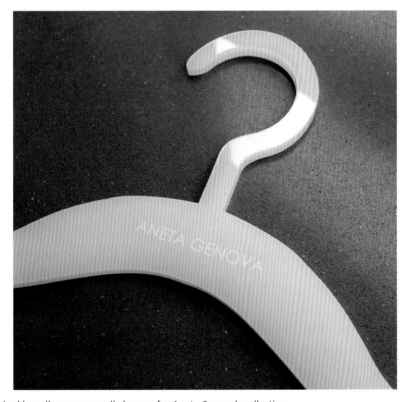

Figure 5.6 Laser etched branding on an acrylic hanger for Aneta Genova's collection.

Types of Fabrication Machines

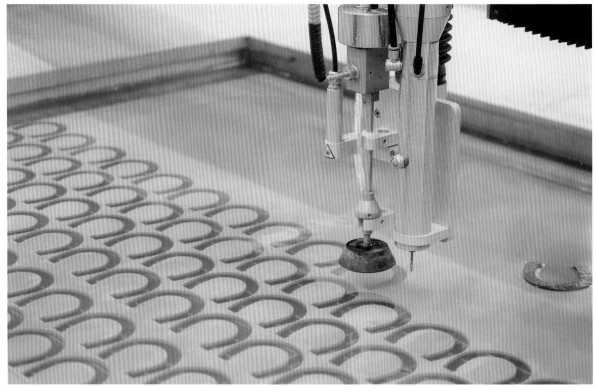

Figure 5.7 CNC cutter/router cuts precise patterns without burning the edges.

is not a good candidate for this process, unless it is adhered to paper, which will hold it in place.

3D Printing

3D printing is an **additive manufacturing (AM)** process. The term originally referred to processes that sequentially deposited material in layers until the entire object was created. Nowadays the term generally includes a variety of **extrusion** and **sintering** techniques that allow 3D models to be turned into physical objects made out of various types of materials beyond ABS plastic, such as various metals, polyamide (nylon), glass filled polyamide, stereo lithography materials (epoxy resins), silver, titanium, steel, wax, photopolymers, and polycarbonate (see Figure 5.8 a, b).

Interlocking 3D printed structures can be used as a textile to create garments and accessories, as seen in Figure 5.1 b. The resulting structures can be designed to have a dense but flexible, or light and airy, look and feel to them. This technique is versatile in terms of developing complex structures with layered effects.

3D printers use the .STL file format, a type of file exported by most 3D modeling software packages. The process of creating 3D modeled objects includes first modeling the object in software, then exporting the file to the .STL format, and then printing the file

 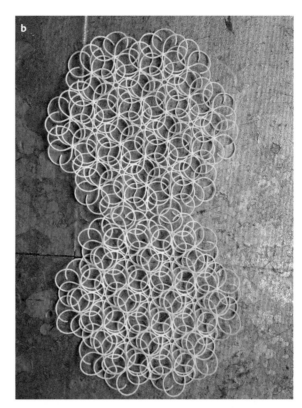

Figure 5.8 a, b 3D printed interlocking ABS plastic generated by Bradley Rothenberg and used by threeASFOUR for their fashion collections.

with the selected printer. 3D printers vary greatly in quality from the less expensive end of the cost spectrum, often used for rapid prototyping, to high-end printers, usually used by service companies that will 3D print files for you. In either case, access to a lower-cost printer is useful for testing and refining a concept before committing to more polished fabrication. Typical software packages used to create 3D models for printing include Rhino, SolidWorks, and AutoCad. Some companies such as Shapeways will even use Google's SketchUp files or the free software TinkerCad.

None of these software environments were created with fashion design in mind. Typically this software comes from the area of product design and architecture or the DIY and Maker communities. Nevertheless, they are being used increasingly for fashion, and designers with the ability to 3D model and print are becoming more in demand, especially in the accessory design industry.

Collaborative Process

Many designers working in this area collaborate with someone who possesses digital fabrication skills. Typically this includes product designers and architects because those fields have engaged with digital fabrication processes longer than others. It is invaluable to learn these fabrication processes

Figure 5.9 The MakerBot® Replicator® is an affordable desktop 3D printer.

in a hands-on manner in order to strengthen your understanding of the specific affordances offered by digital fabrication techniques, but the reality is that, more often than not, in a professional setting there will be defined roles for specific people to construct files for digital fabrication and output the desired form. Increasingly as this type of fabrication is further integrated into the fashion design process, this role may be taken over by individuals trained in fashion, but adept at using software and fabrication as well.

CASE STUDY
BRADLEY ROTHENBERG FOR THREEASFOUR

Bradley Rothenberg and threeASFOUR have been collaborating for a number of years and their relationship, as evidenced by both of their interviews in this chapter, is symbiotic and fruitful. Bradley is an architect by education and his work lies at the intersection of design and technology. In his collaborations with threeASFOUR he is focused on experimenting with generating shapes through digital means and making 3D printed fashion a reality.

The threeASFOUR design trio is famous for experimenting with various technologies and working in collaborative environments. The result of one of these collaborations was a shoe developed for the Spring 2014 fashion show at the Jewish Museum, followed by an exhibition called "Mer Ka Ba." The designs were inspired by the geometry of the different tiling systems in mosques, churches, and synagogues, in an attempt to promote "cross-cultural unity" among

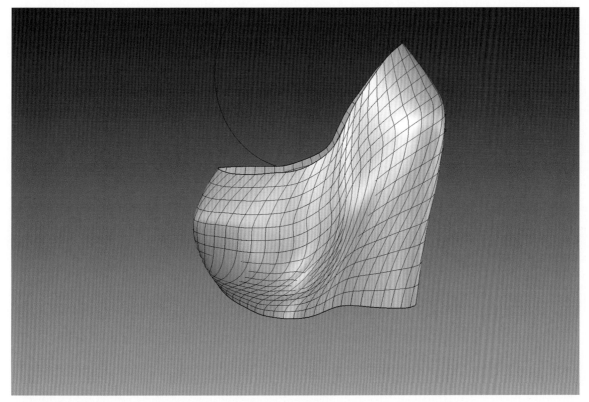

Figure 5.10 3D modeled shoe by Bradley Rothenberg for threeASFOUR.

religions, particularly Judaism, Christianity, and Islam.

This prototype started with a general modeling of the overall shoe shape (Figure 5.10). At this stage the designers determined the heel height, the outsole, and the overall upper shape in 3D modeling software.

In the next stage, the collaborative process includes designing the surface pattern for the upper and creating variations of surface coverage. Considering that this pattern is being applied to a shoe, the designer needs to adapt the shapes to the natural curves of the foot (Figure 5.11). The transition from a flat pattern repeat to one that fits perfectly around the foot is made easier with the help of the 3D modeling software. Traditionally a footwear patternmaker would cover a last with tape and then draw the pattern on it to achieve this effect. This is a cumbersome and labor-intensive process which could never be as accurate as applying the developed pattern with the help of technology.

Once the pattern is applied and the shoe is generated, the designer can rotate and inspect the shoe from various angles within the software to make sure that the scale and placement is correct (Figure 5.12). This is the time to edit the design and apply any changes.

The shoe can also be overlaid on a photo of a real foot in a high heel shoe to assess what it might look like on an actual foot (Figure 5.13). This does not replace a fitting with the actual shoe, but gives a very good idea of the scale and overall shape and fit of the shoe.

Once all changes are finalized, the team moves on to 3D printing of the shoes and fitting on the

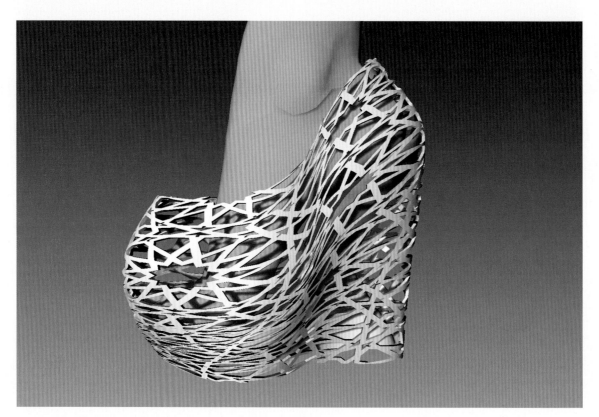

Figure 5.11 3D modeled shoe by Bradley Rothenberg for threeASFOUR.

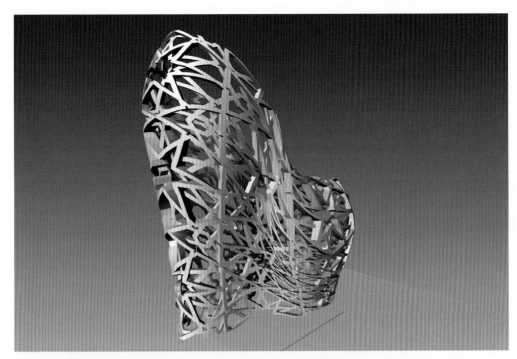

Figure 5.12 3D modeled shoe by Bradley Rothenberg for threeASFOUR.

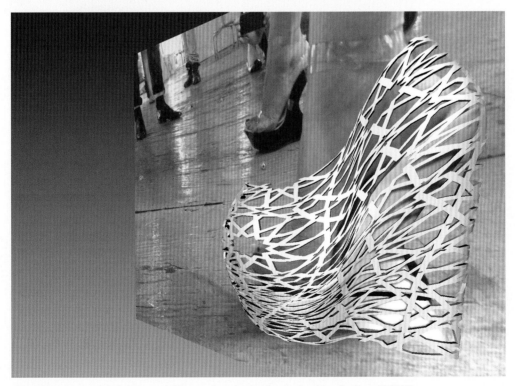

Figure 5.13 3D modeled shoe overlaid on a photo by Bradley Rothenberg for threeASFOUR.

Figure 5.14 Runway shot of the shoes in action (threeASFOUR runway show September 8, 2013).

models to create a cohesive look with the garments in the collection.

At the final stage of creating the shoes, it is important to consider what materials will be best in terms of look and functionality and how long it would take to actually 3D print each particular pair. The partnership with a professional 3D printer, which can handle large volumes of multiple pairs in a short period of time, is integral to the successful completion of the project.

In Figure 5.14 you can see the final product. as shown on the runway for the Spring 2014 collection. The white shoes are created to match the white garments for this particular collection. Keep in mind that depending on the materials you are using, you have diverse options to 3D print in various colors and even layer or mix colors.

Future Trends

The integration of digital fabrication into fashion is part of a larger trend where digital and computational processes impact the design, production, and manufacturing of goods and services. Though the laser cutter, CNC router, and 3D printer are more common digital fabrication tools used in fashion at the moment, technology changes constantly, so this is not guaranteed to remain static. Having a conceptual overview of how digital manufacturing takes a software file and uses that instruction set to numerically control a machine that fabricates physical goods is key to understanding further developments. Indeed many researchers are already talking about nano-fabrication and synthetic biology, in which instruction sets can be applied to either microstructures or sequenced biological chains, resulting in fabricated material. None of this is ready for implementation at this point and remains highly experimental, but these are still nonetheless thrilling possible future choices for designers in this area.

As designers you should consider the possibilities of technology and how they can be incorporated in your design process in more than one way. One method is to use technology as a tool for any stage of the design development, prototyping, and manufacturing processes. A great example of implementing laser cutting to create surface texture is Aimee Kestenberg's Byron bag (Figure 5.15).

Another way to look at technology is as a means to get a head start on the process of creation and generate complex shapes to be used as a "textile," as seen in Figure 5.8. Chapter 6 will provide an overview of programming and new opportunities created by code-based approaches to design. They include using the technology in the iterative and prototyping process as well as for the final output.

Figure 5.15 Inspired by Australia's Byron Bay, this cross body bag by Aimee Kestenberg features a laser cut technique to create a snake scale illusion on smooth surface leather.

CASE STUDY
TRIPTYCH by Tania Ursomarzo

Tania Ursomarzo's 3D printed jewelry is a unique interpretation of the body in motion and the resulting clean, minimal silhouettes are a reflection of her background in architecture. The shapes of the rings are entirely contingent upon the space and form surrounding a specific part of the body—in this case, the right index finger. Her jewelry approach studies the way that material can envelop the finger and highlights the unique spatial and formal condition of this particular body part (Figure 5.16).

For these rings for the right index finger, Tania started the process with several iterations of two-dimensional patterns (Figure 5.17) and created a three-dimensional form, directly engaged with the body at every moment in the process of creation. This resulted was what she calls a laminated "cast" of the body, which captured the unique

Figure 5.16 Tania Ursomarzo's 3D printed ring.

Figure 5.17 Example of two-dimensional laser cut pattern from the initial exploration process.

spatial and formal conditions of this specific body part (Figure 5.18).

This intense experimentation is done manually using trial and error for each stage of the initial process. For her it is integral to the design of these objects that the fabrication—the making—of the work in a sense becomes the work itself, as it is this process that directly shapes (as in, *designs*) the objects. Once she arrives at a form she likes (Figure 5.19), the paper prototype is then 3D scanned to create a 3D model from which a mold can be printed for casting. The finished product is 3D printed in different materials (Figure 5.20): cast bronze from a 3D printed rubber mold, cast blackened steel from a 3D printed rubber mold, photopolymer, elasto plastic, and transparent detail plastic. The weight of each ring changes relative to the type of material, which adds an additional interesting quality to the ring and changes the way that it is worn.

At the end, Tania Ursomarzo developed a technique entirely unique to her own work, which shows how her iterative process impacts the final outcome for each jewelry piece. That process is a blend of traditional and digital methods, which starts with manual exploration and is further aided by technology to create a seamless method where one cannot continue without the other. You can see all of Tania Ursomarzo's work at www.triptychny.com.

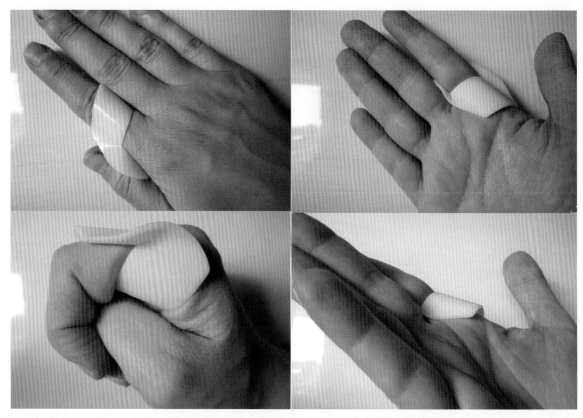

Figure 5.18 These images demonstrate how the ring is shaped around the space and form of the right index finger.

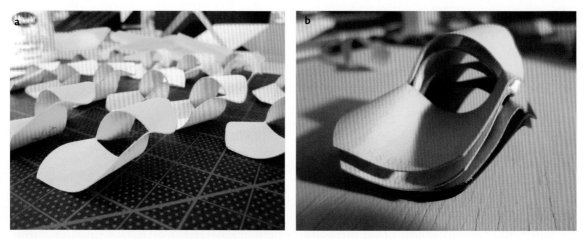

Figure 5.19 a, b Prototypes studying slight variations in the form of the ring, the interchangeability of color due to the laminated method of forming the ring (a), and how the ring's form is stackable (b).

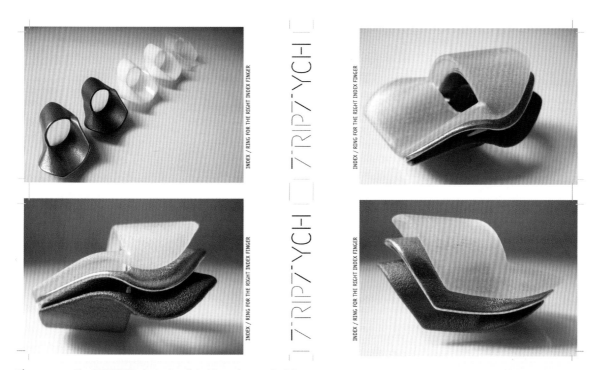

Figure 5.20 The TRIPTYCH rings 3D printed in various materials.

TUTORIAL 1:

LASER CUTTER

Incorporating the use of laser cutter is a fairly straightforward process. While the individual machine you have access to may vary, the general process we outline will be the same.

Materials List

A piece of leather, a sheet of acrylic (or other material approved by the laser cutting facility) you would like to score and/or laser cut, Adobe Illustrator software installed on the computer you will be working with, and a laser cutter.

Step 1: Develop your concept. Remember, laser cutting works particularly well with intricate detailing that is difficult to execute by hand, though this form of digital fabrication can be used for decorative or functional purposes. Laser etching—also called scoring—can be used for development of surface treatment, and it works exceptionally well on leather. In this tutorial we will score a flower design on gold leather.

Step 2 (Figure 5.21): Use an application such as Adobe Illustrator to translate your design into a digital line art file. The Illustrator file in Figure 5.21 represents the design to be etched. Check with the specific laser cutter you have access to. In our case, the laser cutter we are working with requires the file to have an RGB color mode, and it identifies R: 255, G: 000, B: 000 color lines as surface scoring (which we chose to do) and R: 000, G: 000, Blue 255 color lines as cutting. The laser cutter might also have a particular width of line that needs to be designated for each type of line. Ours required a .001-point stroke weight.

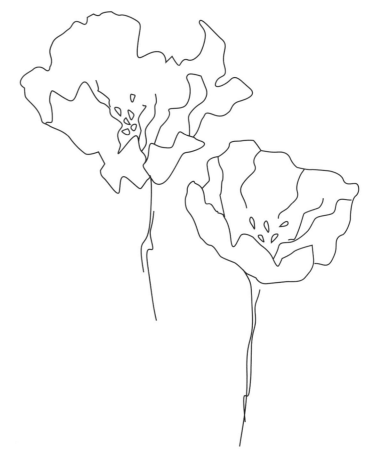

Figure 5.21 Step 2.

174 Chapter 5 Introduction to Digital Fabrication

Figure 5.22 Step 3.

Step 3 (Figure 5.22): Place the material on the machine bed and secure it to the surface if needed. We used weights to keep the corners down and also taped the edges to the rim of the machine bed. Just make sure the weights don't overlap with the work area, and give sufficient space so the laser beam does not come in contact with them. Depending on the material you are using, you should also calibrate the machine to the thickness of your material. There is a difference between working with leather that is ⅛" thick and that which is ¼" thick. The machine needs to be calibrated to the specific height, especially if you are scoring the surface instead of cutting all the way through.

Figure 5.23 Step 5.

Step 4: Using the method specified by your laser cutter, send the file to the print cue for a test run. Select the presets for your specific material. The presets also need to specify what material and what kind of thickness you are working with.

Step 5 (Figure 5.23): Once the laser starts working, watch carefully the progression and stop the machine if you see any discrepancies.

Future Trends

Step 6 (Figure 5.24): Once the process is finished, wait until the machine is fully ventilated and carefully remove the weights and tape. Some items may need additional hand finishing. This will depend on the type of material you are cutting and the end use you have in mind. In some cases, the laser may leave burn marks, which can be hand cleaned.

As another example we also sent an Illustrator artwork file to be cut on acrylic, instead of etched. We created our figure with R: 000, B: 255, G: 000 lines and stroke weight .001 and then followed the same steps as above. The result can be seen in Figure 5.25.

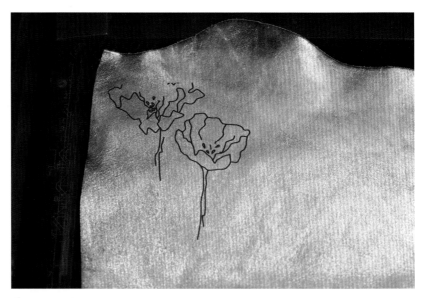

Figure 5.24 Step 6.

Figure 5.25 Laser cut acrylic.

TUTORIAL 2:
3D MODELING / 3D PRINTING

This tutorial will guide you through the steps of 3D modeling, from developing your concept to 3D printing your design. For this tutorial, we are using Jay Padia's jewelry. He developed this collection during his thesis year at Parsons The New School for Design in New York City. He specialized in accessories and developed some accompanying jewelry to complement his line of handbags and shoes.

Materials List

Drawing tools, paper, 3D modeling software, computer, and a 3D printer with filament. If you are working in a 3D lab, you need to check with your particular lab to see if you need to provide your own filament or work directly with them to determine which filament they provide.

You can buy your own filament in various colors and materials, depending on the desired final outcome, but not all 3D printing labs will use the same materials on their machines. Some commonly used filaments include ABS (Acrylonitrile Butadiene Styrene) for hard shapes and nylon for flexible shapes, but there is a great deal of variety you can experiment with.

Figure 5.26 Collaged mood board by Jay Padia.

Step 1 (Figure 5.26): Create a collage for your concept. Designers typically begin defining their ideas through creating a concept collage called a *mood board*. In this example, Jay was inspired by scarabs and lotus flowers for his jewelry designs. He picked the images that best represent the colors, textures, and lines he would like to use and created this mood board to guide the design process. Make sure that you have images rich in information that can give you plenty of conceptual material for the final designs.

Future Trends

Step 2 (Figure 5.27): You can start by drawing some rough shapes on paper before attempting to create a 3D model. For the initial sketches you can just sketch the design lines and then start to finalize how they will be incorporated in the final design.

If you have advanced skills in the 3D modeling software you are using, you can sketch directly within the program (see Figure 5.28 a, b).

Figure 5.27 Step 2.

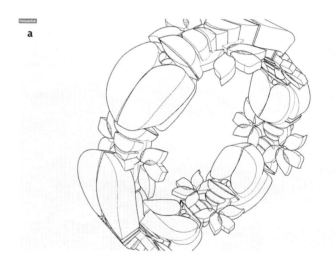

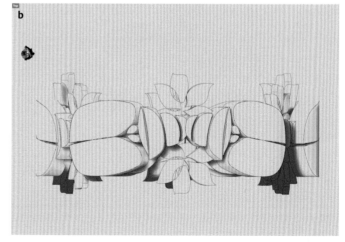

Figure 5.28 a, b Sketches.

178   **Chapter 5 Introduction to Digital Fabrication**

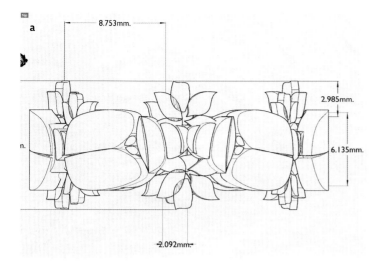

Step 3 (Figure 5.29 a, b): Carefully plan the size of the piece you are designing and make sure it fits the purpose you are using it for.

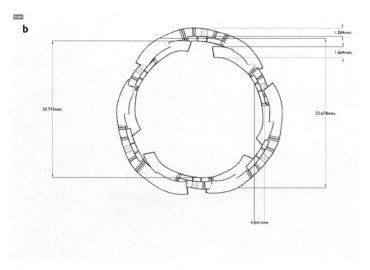

Figure 5.29 a, b Step 3.

Step 4 (Figure 5.30): Render the file and rotate it in various views to see if it looks correct from all angles.

Step 5: Export the file. The software package you will use will likely have the option to export an .STL file. This is a standard file format that most 3D printers accept. This often differs from the standard file format your application will use to save the file. It is important to think about version control and where you save your files. Thinking about this ahead of time will prevent any confusion in case you need to revert to past versions of your work.

A note about orientation: In this particular example, we are using a MakerBot® Replicator® 2X. We needed to prepare the file specifically for this MakerBot. For that purpose we needed to import the file in the MakerBot software and adjust the orientation of the bracelet to lie flat on the printing bed instead of vertical. Since this is an additive process, the 3D printer needs to build support for the extruded parts of the bracelet. In general you want to find the orientation that would need the least amount of support printed around the extruded shapes. By orienting it to lie flat, in this particular case, you can minimize the amount of support that needs to be printed for this irregular shape (and later filed away or otherwise removed).

A note about temperature: The temperature settings for the 3D printer are done in the MakerBot software. Different filaments require different temperature settings on different types of 3D printers. Start with the basic settings of the printer you are using and experiment to find the best setting for each machine.

In our case, we are printing with ABS filament, which is a standard and readily available filament for prototyping in the MakerBot. We set the temperature of the extruder to 230 degrees Fahrenheit (110 degrees Celsius) for the printing bed because these settings have worked best for our particular printer through repeated testing. Test your own printer to discover the optimal settings for your particular case.

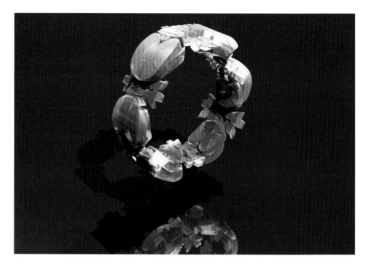

Figure 5.30 Step 4.

Step 6: Check the scale of your piece. Make sure the scale translated correctly after transferring from the 3D modeling software to the 3D printer software. In some cases the scale may shift and need to be adjusted. At this stage you could also decide to print a different size of your prototype, and you can adjust your piece accordingly.

Step 7: Adjust the resolution of the print—in other words, the distance in between each printed layer or the layer height. If you want to print a rapid prototype, you can choose a low resolution and the printer will print faster, but the resulting prototype will be rougher and will have less detail. For a final piece you should choose a high resolution so it prints in fine detail.

Step 8: Send your file to your 3D printer. The MakerBot allows a direct connection to a computer with a USB port, but keep in mind that you need to keep your computer on for the full duration of the printing process, which might take up to a few hours. Some 3D printers do not allow for direct computer connection. In those cases it is better to transfer your file to the SD card that fits your particular 3D printer and input the file that way.

Here we are using a MakerBot Replicator 2X, but the specific machine you use may vary. Printing time on 3D printers may also vary, but tends to take longer than most people initially expect. It is a good idea to build time into your project for multiple prototypes, as the translation from 3D model to printed output can sometimes be very different than expected.

Figure 5.31 Step 9.

Step 9 (Figure 5.31): Once the printer heats up, wait to see that the base layer adheres to the printing bed without any warping, as in Figure 5.31. That will ensure a successful print-out. If you notice any offset or if the extruder is layering filament in unexpected places, exit and restart the printing process. This happens every now and then and it shouldn't worry you.

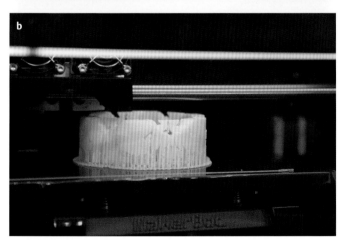

Figure 5.32 a, b Step 10.

Step 10 (Figure 5.32 a, b): Once the printer is in action and is printing correctly, do not move, unplug, or let the computer sleep for any reason. That will cancel the process and you will not be able to resume the printing. You will have to restart with a brand new print from the very beginning.

Step 11 (Figure 5.33): Once the printer is finished, wait until it cools down and do not turn off the machine. For this particular model if you turn off the machine before you have removed the printed object, the bed will rise and your object will be stuck in between the bed and the extruder and will get damaged.

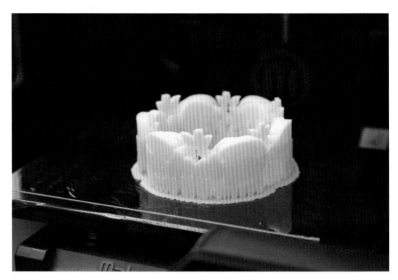

Figure 5.33 Step 11.

Step 12 (Figure 5.34): Remove the print and clean out the support. Figure 5.34 shows an example of the finished printed object as it is being removed from the printer bed. Sometimes, depending on the quality of the printer and the material you are using, you may need to do additional finishing processes to smooth the object.

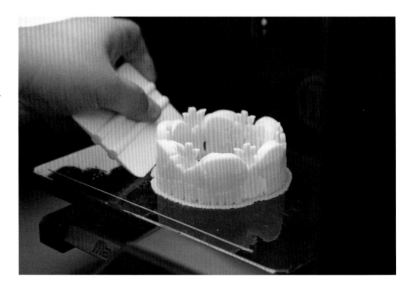

Figure 5.34 Step 12.

INTERVIEW
GABI ASFOUR (FROM threeASFOUR)

ThreeASFOUR is an avant-garde trio who run a fashion label. It was established in 1998 as ASFOUR by Gabi Asfour, Angela Donhauser, Adi Gil, and Kai Khune. Khune left the group to pursue his own label in 2005 and since then the New York City–based label consists of three designers: Gabi, Angela, and Adi.

threeASFOUR won the Ecco Domani Fashion Grant—given to innovative designers in the fashion industry—in 2001. Their work has been acquired by numerous prominent museums throughout the world, including the Victoria and Albert Museum in London. The design trio is well known for utilizing and experimenting with various technologies, including 3D printing and modeling as well as laser cutting their own unique designs.

Figure 5.35 threeASFOUR's Angela Donhauser, Adi Gil, and Gabi Asfour.

AG: Gabi, will you tell us how you got into fashion?

GA: threeASFOUR comes from AsFour. We started in 1998 and we used to be four partners. threeASFOUR came together around 2006/07. From the beginning we were always interested in high tech or, let's say, industrial fabrics, so we worked with fabrics that were used for NASA, for the army. We used to go to industrial fabric shops, and at the same time, we started getting in touch with these German companies that do high tech performance fabrics that are used for heating and cooling. We were interested in these weaves because they were beautiful. And then we discovered *schoeller® textil* in Switzerland, which has couture high tech fabrics. We did a lot of stuff with foam new fabrics that had performance functionalities. The first highlight of technology was digital printing. We used it in a time when people were not really using it. And we also printed digital fabrics based on fractals. We worked on some of these with mathematicians, and we also developed a couple of fabrics by ourselves.

AG: Was there a particular program that you used for this development?

GA: That was done in Photoshop. But the file was about the size of a football field, because the colors were endless. For us that was the beginning of the new tech experimentation, and then we continued into laser cutting. Still using a lot of geometry, too, and that's what led us to realize that we needed mathematicians to work with and let them work on whatever software they need. And then lately it's been 3D printing, which requires Rhino or SolidWorks®.

AG: How do you use these 3D programs? Do you drape the fabric on the body?

GA: You create the volume and shape, and then you figure out how to flatten the pattern. It's basically knowledge of pattern making mixed with what the software can do. An example was a dress that was kind of two balls. And the only way to cut it was to slice it in vertical slices. That dress was 3D printed so we got the flat pattern of the 3D printed dress that we could actually make the fabric from. You can take that 3D shape and flatten the pattern, and when you put it back together it will be exact.

Other things were more like mapping the body. You know, mapping by hand takes ten times longer than mapping with a computer. Right now technology is indispensable. We cannot do most of our designs here without using it. It feels like we would be so backwards if we did not use it.

AG: Would you say that collaborations are key components for the threeASFOUR collections?

GA: This would not be possible without an architect involved. And when you bring in a mathematician, it's even better because you can do more precise things. These are things that I cannot do, but we are very interested in the process and the final outcome. We are working with people who are writing scripts, writing formulas, that allow you to do certain things, and using multiple 3D programs. There's Maya®. There's Rhino. There's Grasshopper. There are surface mapping programs, and lots of programs that allow you to do different things. The only way to do it is collaborate, and it becomes the architect's project as well.

AG: Do you like to work with different people for each collection? Or is it that once you acquire a relationship with people you like to stay with them?

GA: We like to stay with the same. There's Christian Wassmann, who is an architect we are working with, and he likes to create structures. We've done five or six projects with him. He worked with us on the Jewish Museum construction and another one in Holland for a fashion show and we will continue to work with him. We also worked with an intern who came from Harvard last summer and she had some knowledge and was able to accomplish different things, but ultimately I realized the architect we worked with knows so much more. So yes, I'm definitely open to new people, but the people we are already working with are indispensable.

AG: Is there a favorite material you like to work with?

GA: I think cotton for us always repeats throughout the collections. From poplin to twill to canvas, all types of weaves, as well as cotton jersey. It just feels nice, it washes well, and it's easy to wear.

AG: Talking about wearability, how does functionality and wearability fit in with all these incredible technology-derived shapes?

GA: Well, basically you need to create the original thing. You need to let go of wearability at first. But then you can kind of take that and do a version of it. Where you are thinking of somebody sitting, somebody washing it, somebody moving in it. So then it becomes about the requirements of fabric and functionality. Whatever concept was so out there, you can do a version of it that is wearable and functional.

AG: Do you feel that you design certain pieces that are specific for the runway and others for an exhibition?

GA: We have distinct things for exhibitions or more elaborate fashion shows. And then there are garments for selling that can be worn very easily.

AG: And you also work with private clients. Tell us about those collaborations.

GA: Yeah, there are different kinds of private clients. There are a bunch of performers who do exhibitions who want something more for stage. And that is definitely a different approach. That's people like Bjork, Lady Gaga. For us that's a very different approach.

And then there is the evening wear customer who wants something for an event like a gala, and that is a very specific request with a different look and function. And then there's the customer who wants a piece to wear for everyday. So there are three different categories.

AG: Do they come and ask for a specific piece from a collection? Or do they want something custom made?

GA: People actually go back into collections. A lot of them go on the web and say I want this from that season, so sometimes it's not necessarily even from the most recent collection.

AG: That means that you create timeless fashion.

GA: Yeah. It's nice. It feels like whatever we make does not go to waste, because somebody wants it, and we also have the archive pieces here that get rented for movies and commercials and even for events. So there is another side to that. We really have a huge archive in house that is quite organized and taken care of.

AG: What is more important for you: the process of design and working or the final piece?

GA: I think the visualization of something, for me, is interesting. Something that is not possible. Then there's the path to go towards it, and then you are in an unknown territory for the final product, but the most exciting part for me is the path in between. That's where you know you are on the right track but you don't know what the next step is. That's the spark that comes in between. The final, of course, is super exciting. But I'm more interested in the in-between places where you could fail but something else guides you.

AG: Is there a particular collaboration that is very memorable for you?

GA: Yes, there is. We worked with Matthew Barney on pieces that were functional but also super outlandish, and I feel like that type of artist is more of a fitting model for us. One of them was for the performance piece *The Guardian of the Veil*, staged at the Manchester Opera House, and the other was used for his latest movie *River of Fundament* which premiered at BAM (Brooklyn Academy of Music) in February of 2014. I think Matthew Barney is somebody more aligned to the DNA of our design and who really understands our aesthetic. From all the collaborations we've done, his was the one that stood out the most.

AG: For your seasonal collections, do you start with a clear concept? Or do you experiment with materials? Or how does that work?

GA: It depends on the mood. The main thing is that we are three partners. There is some kind of vibe, some sort of feeling in the air, which we discuss. One person says I feel for this, and then another says: "Oh yeah I feel for it." Or they say: "No, I don't feel for this at all." So it becomes a conversation about our instinct in the moment that sort of brings it together. The moment that the three of us are settled, then we are ready to go. Usually we will start sketching, but sometimes it can come from making something with the fabric and the fabric will lead to sketches. So I would say for us there is no really specific set way of working.

AG: Do you guys have different strengths?

GA: Yes, each one of us has different strengths, and we all bring it together. The great thing is that there is an assurance that comes from three places. You are not by yourself with a question mark. With one person, it gets answered once. But with us, if there is a question mark, it gets answered three times, so that's why we can get more confident about what we are putting out. It is a unified vision—otherwise, we don't make it. We really believe we are more like a family than a business. We understand each other from an inner place.

AG: Where and how did you guys meet?

GA: The girls met in Germany. They were a team already. They went to school together in Munich. Then they moved one after the other to New York. And they formed a team here. They were really a stylist team and a fashion design team. I had been in New York for a while, and hadn't found anyone that really I was attracted to. I saw them on the street, and thought: "Wow, this is something that really speaks to me from the heart." Then later we tried to work together, and it worked really well, so we stayed together.

AG: Today there is a lot of talk about wearable tech. In all shapes and forms. Do you feel that is also a part of fashion and what fashion collections are?

GA: I can give you an example. Nike has been using a lot of waterproof zippers, just to be specific, and lots of types of fabrics that they had to create for a certain performance. Those now have become a staple in shows. It's interesting for me that technology is coming in from functionality and it turns into fashion. Things like when you have reflective material for biking and it becomes part of fashion. You see it on the runways. It comes from tech because of need for function.

AG: What about all the data collecting and measuring data? There are a lot of devices. Do you feel that this is also a part of fashion?

GA: The thing is, that whatever they make for these devices, they need to get a designer to make them, because it needs to look good. And if it fits on the body and is designed for the body, then right away it becomes fashion. There's no way for it not to become fashion.

AG: What kind of advice do you have for young designers who are very much into fashion design and are hesitant to approach technology?

GA: There is no way that companies like Hermes or Chanel will survive unless they use technology because you can do so much more with technology. I urge any young designer to at least try to figure out a way to use technology. For fashion specifically, 3D printing is a tremendous advancement. Usually there is a separation between the fashion designer and the architect, so it needs to have some sort of merging together. Same with laser cutting, because geometry comes into it, or any other technology—it brings a need for collaboration. This is the way that any fashion designer will work in the future. They will have to collaborate.

AG: What technology are you working with right now?

GA: We are working with projection mapping, and we are working on a few films that have a lot of heavy After Effects requirements. So we just did a film last year with a 3D animated fractal, where the environment is pure geometry, and that was extremely effective, but there are plans of projection mapping where you have somebody moving alongside it. We are in touch with Intel, which sent us a device, a projection mapping apparatus, which we are testing right now, so hopefully something will come from that. And I'm excited for the future.

AG: Any last words?

GA: Just keep in mind that 3D printing is also happening through collaboration. The company that printed our last two or three fabrics is Materialize. They need us to experiment with their technology, so they can gain a foot in the fashion industry, and we need them to support that sort of thing and to develop it. It's a pretty nice collaboration. And now we are in touch with Stratasys, which also offered us another collaboration, as well as Autodesk, who created Maya®. These are large technology companies that are developing their products who want fashion designers like us to work with so that they can keep innovating.

INTERVIEW
BRADLEY ROTHENBERG

Bradley Rothenberg is an architect and designer who has worked in various architectural environments around the world, including a tenure with Acconci Studio, which oversaw the development and execution of pavilions and multimedia installations in Italy, Hong Kong, and Holland. He has collaborated with artists and designers such as Ai Wei Wei, ThreeASFOUR, and Katie Gallagher. His independent work has been exhibited at the 2009 Design Miami and the 2010 Hong Kong-Shenzen Biennial.

His work exists at the intersection of design and technology. Bradley Rothenberg is focused on making 3D printed fashion a viable and accessible reality. With the goal to unlock 3D printing's potential to change design and manufacturing, the studio explores computation as a method to generate objects that could not be made in other ways. Bradley Rothenberg has worked with some of the most innovative fashion brands to incorporate 3D printing into their lines, while introducing them to the capabilities this new technology affords. He regularly collaborates with threeASFOUR.

AG: You are an architect by education and by trade. How did you transition into the world of fashion?

BR: While working with Vito Acconci, we always spoke about fashion being the first architecture, as it is the first layer of protection for body from the environment. Also, studying architecture, I became more interested in geometric systems and processes to grow and make form, rather than the actual buildings themselves. When I first started collaborating with fashion designers, I realized there was a huge opportunity to apply the form making processes on a totally different scale from buildings, and finally, being introduced to Selective Laser Sintering, there was a method to actually make in real life the geometry I could generate on the computer.

As I continued to be introduced to the fashion industry, I started to see that there was a disconnect between designers and textiles, it seemed like the designers would go into a store and pick out different fabrics and then work from there, rather than actually being able to generate the fabric themselves. This was where new technologies like 3D printing could start to give more control to the designers and allow them to develop custom textiles—allowing for the weave itself to make the textile.

Figure 5.36 Bradley Rothenberg in his studio.

Future Trends

AG: Tell us about your design studio and the kind of projects you work on.

BR: Cellular textiles and methods of optimization. We are focused on performance materials, which include the customization of shapes to other shapes, and the development of what we are calling cellular materials. Materials that are made from variable units assembled together with specific properties in specific areas. Also, we are consulting with fashion designers and other companies to develop projects for them using these high performance materials, which act as use cases for them. Our overall focus is in the development of software that allows others to develop these custom materials that are now possible because of advanced manufacturing methods.

AG: What is your design philosophy?

BR: Process and iteration.

AG: Tell us about your design process. Do you start with inspiration or the methodology and technique first?

BR: We never start with inspiration. That seems too much like artist thinking, almost as if the project comes from somewhere else, or some higher place. Rather, we are interested in developing methods specific to the project, working out a new technique to allow us to do something that hasn't been produced before.

AG: How and why did you start working with 3D printing? What was your first creative investigation into it?

BR: I was introduced to 3D printing while at Pratt in 2005. I worked in the digital fabrication lab when we set up the first 3D printer in the architecture school. By today's standards, zCorp Powder based printing is old technology, though at the time it was exciting in that it was really the first time I could see a method of actually making the complex 3D models I was so interested in being able to generate. My first project utilizing 3D printing was actually an architectural pavilion in school, during a studio w/Evan Douglis. The direction of the studio was to research how new manufacturing methods, specifically 3D printing, could allow us to make form that could not be made with other methods.

AG: How has technology influenced your career?

BR: It allows things that were not possible to now be possible. I see technology as a means, not necessarily an end; technology opens up a whole new range of possibilities, rather than solving one specific problem. At the same time, technology doesn't usually come first—rather, we think of something we want, like complex interlocking textiles, then find or develop, if it doesn't exist, the technology to make that possible. Then that technology opens up a whole range of stuff we didn't think of, and hopefully can influence others in ways we didn't initially consider.

AG: What is your vision for 3D printing and its future?

BR: I hope to see the technology continue to increase in resolution and be able to print smaller and smaller, so we can make more and more comfortable materials. The future vision is to see printing on a molecular level, so we can actually build up materials molecule by molecule to have full control over the actual material that is being made. Also, I think 3D printing can redefine what is now a pretty stagnant retail experience. Imagine going to a store and now being able to pick the exact unique piece of clothing or accessory customized to your body. Now the manufacturing of that part can be localized to each store, so each store in a way is like a mini-factory, constantly pumping out new custom items.

AG: What technical innovations would you like to see in the world of fashion?

BR: Highly customized seamless clothing. I think there is a huge opportunity for technical innovations for the design side of the world of fashion. Some fashion designers are still using age-old methods to cut and drape patterns, and materials are still made with knitting and weaving methods. The computer is just starting to be incorporated in fashion designers' processes and it will have huge implications for how clothing is made. On the consumer side, I think retail stores have to completely rethink what they are. Rather than just a place to buy clothing, the store has to envelop the consumer in the experience of the brand. Technology can help the store become interactive, feeding and receiving information from the consumer.

AG: What advice do you have for young fashion designers?

BR: Take as many computer classes as possible while in school.

For Review

1. How do you define digital fabrication?
2. What are the various fabrication machines a fashion designer can use for rapid prototyping?
3. What is the difference between subtractive and additive fabrication processes?
4. What software do you need to master in order to work with each one of these tools?
5. What are some of the best ways to utilize a laser cutter for the fashion/accessory design process?
6. When is the CNC router preferred over a laser cutter in the fashion industry?
7. What kind of files do you need for the printing process and what software produces those files?
8. Why do fashion designers need to collaborate with technologists?
9. What are some of the newest technologies using instruction sets for fabricated materials?

For Discussion

1. What is your current experience with using technology for prototyping or fabrication and what challenges have you experienced?
2. Which components in handbags or in footwear do you envision can use the 3D printing rapid prototyping and why?
3. What kind of methods or technology would you use to generate your own fabrics?

Books for Further Reading

Gramazio, Fabio, Matthias Kohler, and Silke Langenberg. *Fabricate: Negotiating Design and Making*. Zurich: Gta-Verlag, 2014.

Hoskins, Stephen. *3D Printing for Artists, Designers and Makers: Technology Crossing Art and Industry*. London: Bloomsbury, 2014.

Johnston, Lucy. *Digital Handmade Craftsmanship in the New Industrial Revolution*. London: Thames & Hudson, 2015.

Kaziunas. Anna. Make: *3D Printing*. N.p.: Maker Media, 2013.

Lipson, Hod, and Melba Kurman. *Fabricated: The New World of 3D Printing*. N.p.: New York: Wiley, 2013.

Warnier, Claire, Dries Verbruggen, Sven Ehmann, and Robert Klanten. *Printing Things: Visions and Essentials for 3d Printing*. Berlin: Gestalten, 2014.

Key Terms

Additive fabrication: A fabrication process that creates an object through adding small amounts of a material until an object emerges.

Additive manufacturing (AM): Processes that sequentially deposit material in layers until the entire object is created.

CNC routers: A computer-controlled machine for cutting various hard and soft materials, such as wood, composites, aluminum, steel, plastics, foams, fabrics, leather, and other materials.

Digital fabrication: A fabrication process in which a machine is controlled by a computer.

Extrusion: A process used to create objects of a fixed cross-sectional profile in which a material is pushed or pulled through a die of the desired cross-section. In this book, 3D printing refers to extrusion-type printing: raw material being melted and formed into a continuous 3D shape as defined by the design in the input file.

Laser cutting: This process uses lasers in a subtractive digital fabrication process.

Sintering: The process of compacting and forming a solid mass by heating a material without melting it to the point of liquefaction and depositing it at points in space defined by a pre-designed 3D model, binding the material together to create a solid structure.

Subtractive fabrication: A fabrication process in which machines chisel or cut away at a block or sheet of material.

6
Introduction to Code

This chapter focuses on the flourishing activity around **code**, programming and computation as a method for generating expressive design patterns, shapes, and prints. Software, driven by code, is what underlies many of the systems in contemporary society and has become an increasing force within the design world.

This chapter outlines some of the opportunities for designers to integrate programmatic thinking in their work. Working with code might appear to be a dreadful and even impossible task, but understanding the code-based process will give you an opportunity to create complex designs that otherwise would not be possible using existing general graphics and multimedia software and to collaborate effectively with programmers if needed. The beauty of code-based design is that it can be applied to many parts of the design process, from surface treatment and print design to creative pattern making.

After reading this chapter you will:

- Understand the difference between code as a process for design vs. code for task-based and functional applications.

- Be able to distinguish between computer scientists and research groups vs. creative coders who approach code from a more poetic or expressive perspective.

- Understand how code and software connect all the disparate systems we have and how code allows these systems to "talk" to each other.

- Be introduced to the Processing platform and understand how to use coding as inspiration to generate new form.

- Know how to export code-based designs via digital fabrication to create new forms.

- Have insight into the collaborative process between designers and technologists in order to design with code.

Code

In recent years the language of computation and its use for expressive purpose has grown enormously. Software drives nearly everything in contemporary society, from the latest smartphone app to stock exchange trades. Code and computation are an enmeshed part of contemporary living, creating opportunities for fashion to incorporate these influences in new and surprising directions. For designers, an understanding of code, conceptually and practically, allows the designer greater control and precision in expressing ideas from concept to production.

Code is a system of rules that converts information into another form or representation for communication through a channel or for storage in a medium. Code is first and foremost a mental construct, wherein instructions or a sequence of actions can be specified. This represents a shift within the field of fashion, which has much of its tradition and ways of making rooted in either handcrafted or industrial processes. In the "information age" software undergirds our daily lives, providing the connectivity, processing, storage, and retrieval necessary for everyday systems to function. As fashion has always expressed the values and aesthetics of its time, there is an opportunity for designers to convey something about how code is impacting society. This opportunity comes not in just referring to code in one's design process, but actually using or writing code as part of the creative process.

In previous chapters we discussed many new approaches for creating fashion that require code to operate. For example, electronic development platforms require software to provide instructions as to how a garment or an accessory should perform. In other words the computer needs an instruction set to tell it what to do. Likewise, with digital fabrication, it is the computer using code, which directs the cuts of a laser cutter or the filament deposit of a 3D printer.

Ada Lovelace is often referred to as the world's first computer programmer, having in the mid-1800s outlined what is now recognized as an algorithm, a detailed step-by-step set of operations to be performed. Though modern computers as we know them did not exist at that time, the conceptual basis for a computer program was established through her notes on Menabrea's paper on Charles Babbage's Analytical Engine. (Charles Babbage is credited as the first to conceive of the possibility of a programmable computer.) The notion of an instruction set that can be followed in detailed, step-wise fashion is at the core of how software functions.

For designers there is an opportunity to use code and its operation, either conceptually or in an applied manner, to explore new methods for generating creative inspiration, as well as methods for clothing design and construction. While it is not necessary to work directly with code to create relevant work in this area, an understanding of programming is a powerful skill set to possess. General graphics and multimedia software have a high creative ceiling, but are still limited by the constraints established by the creator of the software application. There are artists and designers who work with computation in ways that move beyond manipulation of effects and settings in software applications to design entirely using code. This type of design process is increasingly at the forefront of innovation and young designers can benefit greatly from familiarizing themselves with this area of creative production.

Creative Coding

Creative coding is a term used to describe a growing community of artists and designers who use programming to create expressive artifacts. Typically found at the intersection of fine art, advertising, and design, creative coders tend to work with open source development platforms and participate in a robust community of developers who share and build upon each other's code and contributions. Unlike traditional computer science, which is a wider field of study and more broadly focused on theoretical and practical applications of computers, creative coding tends to emphasize aesthetic and social experiences, and is typically experienced in art and entertainment venues. Practitioners tend to come from a wide range of backgrounds, many of them from studies in art and design, or they engage in dual studies in both engineering or computer science and the visual arts.

Development environments such as **Processing**, **Cinder**, and **openFrameworks** provide coding platforms that are oriented towards programmers who come from a non-programming background and aspire to write software to create visually stimulating designs for artistic expressions. While many creative coders might have a traditional computer science degree education, often the approach and sensibility of creative coding departs from traditional engineering values of optimization and efficiency. Creative coding tends to focus more on an aesthetic outcome or experience, with the use of code likened to a paintbrush or poetry.

Many in the creative coding community find general graphics software and media formats restrictive, preferring instead to build their own software to individualized specifications rather than using commercial software products currently offered. For fashion designers creative coding can be used as a primary tool for direct creative expression, using the native platform to develop form, shape, and print patterns.

Using Code

There are many ways you can incorporate code into the design of garments and accessories. How and what you implement will depend on your knowledge or desire to learn as well as the type of collaborations you engage in. Keep in mind that knowing what you would like to achieve is a valuable part of the process and will allow you to seek the appropriate software platform or collaborator who can help you get the result you want. Through digital experimentation, you can create not just surface patterns, but also reach another level of creative pattern making. Some primary ways to work with code include the generation of pattern for print, surface treatment, and shape and form.

Patterns for Print and Surface Treatments

Visual patterns for use on textiles, or as inspiration in the form of texture and technique, can be generated through software. For example, the works of Cait Reas draw inspiration from the art of Casey Reas (Figure 6.1), an artist who works primarily with code to create dynamic and evolving visual artifacts. In this case, the output of a software process is the starting point for the design process. The final garment does not necessarily need to have any functioning code or an embedded responsive hardware display. Instead the formal elements of the code are worked into the core concept and the output is a traditional garment, using visual language that has its roots in procedural logic.

Many software artists, such as Casey Reas, Joshua Davis, Jeremy Rotsztain (see Figure 6.2) and LIA (see Figure 6.3), use code to create stunning images, which can be incorporated into the fashion design process. In the same way art has always influenced fashion, the work of artists who work primarily in code can be a powerful source of inspiration, made more meaningful when the underlying methods of creating the work are understood as an essential and integral part of the produced visual artifact. Artist collaborations are common in the fashion world, and as the prominence of software art grows every year, it has become more common for fashion designers to collaborate with artists who work in this area.

Various methods of collaborative creation can be implemented throughout the design process. Some designers may choose to work with an artist from the onset of the concept stage and be a part of the image creation itself; others may choose to let the artist create the imagery and then use it for a print, a surface treatment, or creative pattern interpretation. In some cases the method of creating the digital imagery can be the inspiration itself

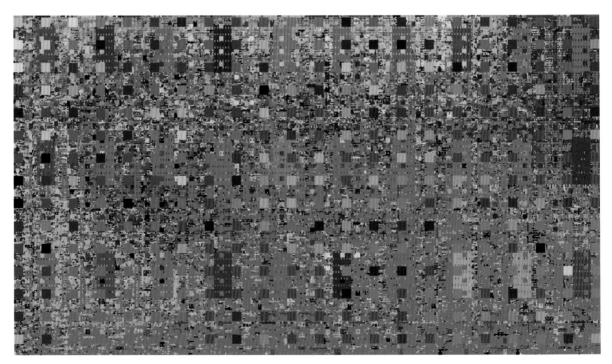

Figure 6.1 "Pfft!" Digital artwork created by a series of coded instructions that mix video content written by Casey Reas.

Figure 6.2 Action painting still from a video by Jeremy Rotsztain. Action Painting is a series of still and animated digital paintings composed using moving visual elements from Hollywood action flicks as compositional material to create works in the style of abstract expressionist painters such as Jackson Pollock. Those elements include explosions (from *Armageddon* and *Independence Day*), fistfights (from *Fight Club* and *Rocky*), car chases (from *Bullitt* and *Bourne Identity*), and gun shots (from *Rambo* and *Terminator 2*), which become compositional material to create the works.

Figure 6.3 Software art by LIA. Live Visualization from a Gustav Mahler concert in Vienna, Austria 2011.

for the fashion concept. What is unique in such a partnership is the variety of creative language and methods used; each one can produce a unique collaborative outcome. As a designer you may already be used to selecting a color palette or an image to work with, but such collaboration can push you to explore a new direction for development or creative process, which will produce new ideas. Do not be limited just to the imagery itself, but explore a new approach of creative thinking and 3D form exploration and design.

Shape and Form

Code can be used to generate 2D and 3D shapes and forms that can be used either as a starting point for inspiration or as the shape and form of the garment or accessory itself. In the previous chapter digital fabrication was defined as a manufacturing process in which a fabrication machine is controlled by a computer. While 3D forms and 2D shapes can be created through the use of modeling software, there are also designers who utilize code to generate original silhouettes, which can be replicated in multiples or otherwise connected so they can be used as a textile (Figure 6.4 shows Bradley Rothenberg's 3D materials) or even the final garment or accessory (see Figure 6.5).

For example, the system created by Jessica Rosenkrantz and Jesse Louis-Rosenberg creates complex, foldable forms, which can build an accessory like the one in Figure 6.5, composed of articulated modules. By combining computational geometry techniques with rigid body physics and customization, this system provides a way to

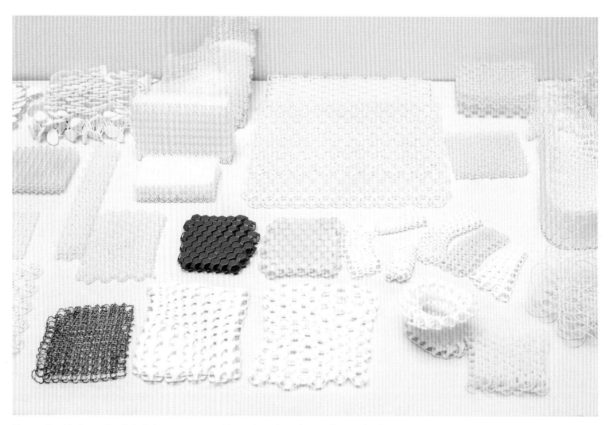

Figure 6.4 Various 3D printed shapes generated by code written by Bradley Rothenberg.

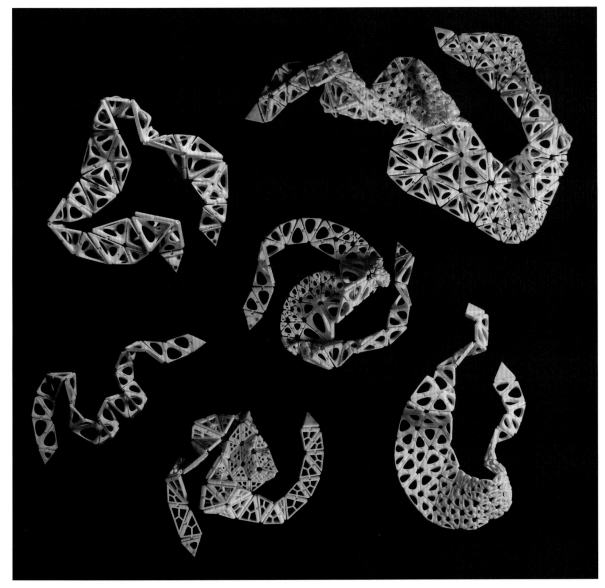

Figure 6.5 Kinematics necklaces created by Nervous System.

turn any three-dimensional shape into a flexible structure using 3D printing. That means that any object can be simulated and compressed down for 3D printing through intricately patterned and interwoven components. In this example we can see that the necklaces are so flexible that they can conform to the body with ease. Each individual component is rigid, but combined all components behave as a continuous fabric. Even though the complete "fabric" is made of many distinct fragments, the final design requires no assembly thanks to the 3D printing technology, which allows complex structures to come ready straight out of the machine—evident in these examples.

Programming Concepts

While there are many different programming languages that exist, in general there are three characteristics of a programming language that are often shared:

1. Sequence of commands: Programs comprise an instruction set that is sequenced in a particular order. In the same way assembling a garment requires a pattern to be cut and sewn according to predetermined steps, programs rely on a specific sequence of commands to follow in order to run. When writing software it is important to have these steps clearly identified in order to write code that will work without **bugs**, a term programmers use for errors.
2. **Conditional structures**: Programs require certain conditions to be met in order to proceed along a predetermined sequence. The pattern-making process uses labels and notches to indicate how pieces of fabric should go together. Evaluating whether these pieces should be sewn together requires an assessment that can be summed up in a conditional statement. For example, if a seamstress is sewing a sleeve, she will ask herself: "Do the notches on this sleeve align? If yes, continue to sew them together. If not, stop and find the proper sleeve pattern piece!"
3. **Looping structures**: These are processes that repeat themselves in a sequence of steps until a certain condition is true or as long as a certain condition remains true. An example of this from garment construction might be a particular type of embellishment along a hem, which repeats until the edge of a garment is reached or repeats for only a certain number of times.

Programming concepts are logic structures that dictate the sequence of events in a program. Programming languages are structured ways of expressing those concepts in a form that computers can understand.

There are many programming languages in existence. Each language expresses programming logic in a specific way. Like spoken human language, some have similarities to each other and others are radically different. Again like spoken language, each one has its own rules of grammar and syntax. The language a computer program is written in is different from the development environment. The language is a possible set of instructions; the development environment is where writing in that language takes place.

Processing

Within the creative coding community, there are a wide range of programming platforms that practitioners choose to work with. Like any tool, the choice of a computer language and environment comes with particular affordances. Ben Fry and Casey Reas created the Processing language and IDE (Integrated Development Environment) in 2001 to teach computer-programming fundamentals in a visual context. Since then the environment has grown into a robust community, with artists, designers, researchers, and hobbyists using processing for a wide range of purposes, from the personal to the commercial.

Processing is an open source development platform, freely available for download on the Internet. Over the years Processing users have added to the functionality of the platform through contributed code examples, libraries, and documentation. For designers who are interested in getting familiar with computer programming, Processing provides an excellent opportunity to learn some of the basics in a relatively easy and friendly manner. The processing website hosts a range of tutorials and examples, as well as links to various books that have been written on the topics. There is also a robust online forum where people can post questions and seek help. In addition, with the rising popularity of Processing there are short courses and workshops (often locally organized by city) that are offered to help people in a more hands-on manner.

Other popular environments for creative coding include Open Frameworks and Cinder, though neither is as friendly to novice programmers. Still, when collaborating with programmers you may find that they may work in a different environment not mentioned here, as programming languages evolve. Regardless of development environment or language understanding, basic programming concepts and logic will help you when working on projects with a collaborative team or as a starting point for your own practice employing code in your work.

TUTORIAL: 1

LEARNING CODE

Designers who are interested in learning to code have a great deal of choices available to them nowadays. Some of the more popular coding platforms for creative coders are Processing, OpenFrameworks, and Cinder. Programming platforms will change over time, with new environments being introduced and others eventually receding in prominence. This tutorial will take you through the process of getting to know the Processing platform. This is only a rudimentary introduction to programming and should be seen as a starting point for further learning.

We chose Processing as a beginner-friendly environment to learn about programming. While other environments may have particular features that are valuable, the creators of Processing—Casey Reas and Ben Fry—developed this environment with the express purpose of making the process of learning to program more attainable for a wider population.

Processing can be useful for generating patterns for prints on textiles and leather or generating visual assets that might inspire a collection. Alternatively, learning the basics of code may be a knowledge base acquired in order to facilitate better collaboration with programmers.

There are numerous resources online and in print available on programming in Processing. Visit the Processing website and the various books on the topic that are showcased there. They go into greater depth and detail on the subject of programming and generating visual expressive artifacts in code. In this tutorial we will take you through the process of creating a simple program in Processing. The code will produce rudimentary visual design. The tutorial applies the output of the provided code to a croqui, but you can take the outcome to use in your design process in any way you like.

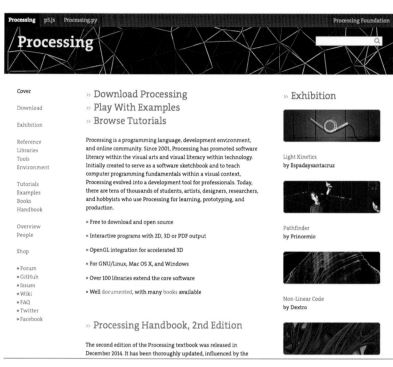

Figure 6.6 The Processing.org website landing page.

Step 1 (Figure 6.6): Visit the Processing website: https://processing.org/. The Processing website provides a range of resources and links to information about the programming environment, the community of volunteer developers who maintain and add to the environment, and references such as tutorials, code examples, and sample projects.

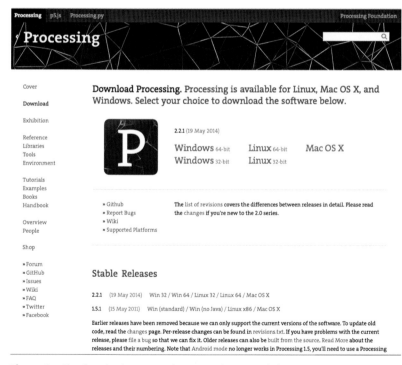

Figure 6.7 The downloads page on the Processing.org website.

Step 2 (Figure 6.7): Click on the download link and select the right version for your operating system. When the program has downloaded completely, install the software on your computer by double clicking the .zip file on your Mac or PC.

- In Windows, drag the folder inside the zip file to a new location on your machine, typically wherever you store your applications. Launch the processing .exe file.

- On a Mac, you will see a Processing icon. Move this icon to the Applications folder of your computer.

Step 3 (Figure 6.8): Double click to launch the application. When the program launches, you will see the application window. (If you do not, refer to the troubleshooting information on the Processing site.) In the upper left corner of the application window, you will see a set of controls. Those controls are replicated in the drop down application menu as well.

Step 4: Copy and paste or carefully rewrite this code exactly as it is into the application window. The following is based on the "Getting Started with Processing" tutorial on the Processing.org website,[1] which is in turn modified from the book *Make: Getting Started with Processing* by Caesy Reas and Ben Fry.[2] (There are many excellent tutorials introducing novices to programming in this section of the website.) The following code is modified from that tutorial. (While code is often freely available for use in the open source community, it is very important to give credit to your sources and to clearly identify when you have used or been helped by another person's code.)

Figure 6.8 The application window in Processing.

```
void setup() {
  size(480, 320);
}
void draw() {
  if (mousePressed) {
    fill(0);
  } else {
    fill(255);
  }
  rect(mouseX, mouseY, 60, 60);
}
```

1. Processing.org. "Getting Started with Processing," (Tutorial) accessed June 19, 2015, https://processing.org/tutorials/gettingstarted/
2. Casey Reas and Ben Fry, *Make: Getting Started with Processing* (Maker Media, 2010).

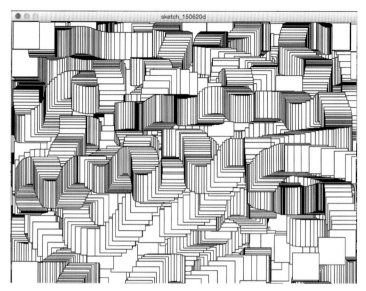

Figure 6.9 The output of the code in the "Getting Started with Processing" tutorial.

Step 5 (Figure 6.9): Hit the "Run" button and see the results. What you will see is a white square on a 480 by 320 pixel canvas. The square follows your mouse movements. When you click your mouse and hold down on the button, the square turns black. When you unclick, the square returns to white. The square continuously redraws and leaves visual traces as it moves.

Step 6: Once you have the code working, you can change the parameters of the code to see visual changes take place. For example, you can change the size of the sketch window from 480 by 320 to 480 by 480 by tweaking the code on the second line. Simply change the value from 320 to 480 like this:

```
void setup() {
  size(480, 480);
}
void draw() {
  if (mousePressed) {
    fill(0);
  } else {
    fill(255);
  }
  rect(mouseX, mouseY, 60, 60);
}
```

Step 7 (Figure 6.10): Hit the "Run" button and see the results. As you can see, the pixel size of the window has changed to the size set in the code. You have effectively made a change to the code and seen the results take place. Keep experimenting with the code until you are happy with the visual material you've generated. Keep enlarging the window if you need a larger piece of art.

Now let's go through the code line by line to see what happens as the program runs. There are two main parts to the code set up as two separate functions: void setup() and void draw(). The void setup() function executes whatever is inside the brackets {} that follow. In this case, "size(480, 480)" refers to the dimensions of the display window in units of pixels. The first value represents width, and the second one height.

Figure 6.10 The results of the edited code—transformed window size.

Figure 6.11 Fashion croquis utilizing the resulting processing print-out for garment designs.

The function "void draw()" refers to the main area of program execution. The program loops through the content between the brackets {} after "void(draw)."

In this case, we first have a conditional statement, "if(mousePressed)," which basically looks for whether or not the mouse attached to the computer (or track pad for a laptop) is pressed. The fill() function determines the fill color of shapes. This gets declared before a shape is drawn. When the mouse is pressed, the fill color is set to 0, which is black. (Original web colors are represented by a number between 0 and 255.) When the mouse is not pressed, the fill is set to 255, which is white. Then the rect() function is called, drawing a rectangle. The rect() function has parameters, which set x position, y position, width, and height. In the example code, the position of the square is tied to the mouse position (mouseX and mouseY), and the color is determined by whether or not the mouse is pressed.

Further explanation of the many functions available in Processing can be found in the "Reference" section of the website (Processing.org), along with proper syntax (how to format the code to achieve your desired result).

Step 8 (Figure 6.11): Print the developed pattern and apply it onto your own fashion design sketches. In this example, we applied the developed artwork onto already-sketched fashion croquis. Collaging the prints in various ways starts the concept development for a fashion collection. Try gluing flat pieces as well as some three dimensional forms made by pleating or otherwise folding the paper. Vary the scale of the prints to achieve different effects. Once the pattern is applied, you can also add color or other textures and materials to develop garments and accessories.

As you begin to experiment with programming on your own, you will find a robust community of practice offering tutorials, forums, code examples, and code libraries. The Processing website should be your first step for research and experimentation. After you go through the "Tutorials" section of the website and learn the basics of working with Processing, you should explore the "Examples" section and look through the existing projects. They are divided by form, data, images, color, typography, and so on, and you can find an array of visually exciting projects to get inspired by.

Keep in mind that this is open source software and the source code is available to be studied, changed, and distributed to anyone and for any purpose. The published projects contain contact information about the developer who created that particular project and information about how to properly credit their work. It is acceptable to copy and reuse code that has been written by another person, provided attribution is properly acknowledged. Code written for open source distribution typically has what is called a "header" at the start of the code that explains how the original author would like credit to be given for their contribution. It is very important to abide by those stipulations and, in exchange, open source contributors are often very generous with how their work is replicated and disseminated.

Look for informal programming workshops and courses, available in varying lengths and areas of specialization. Attending these learning opportunities can be invaluable in terms of meeting like-minded potential collaborators and advancing your own knowledge of the field. This will help you find new ways to apply programmatic thinking to your projects, whether you become proficient in programming or simply gain enough of an understanding to conceptually keep up with members of a collaborative team.

CASE STUDY
FASHION FIANCHETTOS

Otto Von Busch is a trained fashion designer who works with interaction design methods. In the workshop Fashion Fianchettos, Von Busch developed a system for draped algebraic topology.

A fianchetto is a tactical move in chess, especially popular in "hypermodern" chess, which aims at using the pawns to clear path for movement along the bishop's diagonal towards the center of the board. This type of move can be a fruitful metaphor for rethinking the way fashion is disseminated throughout fashion ecologies.

Envisioned as a way in which to transform fashion into "software" through encoding draping technique, the workshop provided participants with a handful of safety pins and an oversized t-shirt. This approach took the fashion design process as a set of almost mathematical rules or codes through which they could "reprogram" the clothes with simple changes in the execution of code. The result was a platform, with the t-shirt as basic "hardware" and the notation scheme as the basis for "software." The programming language was named *Fashing*, and it uses predefined commands followed by coordinates on the t-shirt.

Figure 6.12 Participants of the Fashion Fianchettos workshop use an oversized t-shirt and safety pins to re-think the "codes" of garment construction.

Figure 6.13 The grid pattern visible on the oversized t-shirt is used as a set of coordinates to reconFigure the garment based on the written code.

Below are some examples of the commands (with a description of each after two slashes //):

- garment // status or type of garment
- grid(x/y) // size of grid coordinates, if not standard
- wear(x) // how to wear garment (backwards, etc.)
- invert // turn garment/section inside out
- connect(xy, xy) // function; connected coordinates
- switch(x) // turn garment/section around
- foldDouble(xy, xy) // fold and pin through all layers
- stretch(xy, xy) // stretch fabric between coordinates
- repeat // repeat commands (such as many pleats)

In this workshop participants merged draping, a method of precise step-by-step procedures, with a coordinate-based system laid over the t-shirt. Commands authored by participants and executed on the garments determined which coordinates are connected via the safety pins. The result is a series of points connected on the garment that transform the form and appearance, allowing the t-shirt to be shaped in new ways without damaging it. Each step can be undone or remade, allowing the garment to "reset" back to its basic shape, allowing for indefinite updates to the garment.

A sample Fashing program is presented below. The structure of computer code is used to describe the precise method by which participants should reconFigure their "hardware," the oversize t-shirt. At the beginning is a list of example commands, much like a "language reference" for programming languages, and the code specifies how the coordinates in the system should be connected. You will see how basic code conventions apply to physical objects (the t-shirt, and the direction to "wear as normal"), yet are programmatically sequenced.

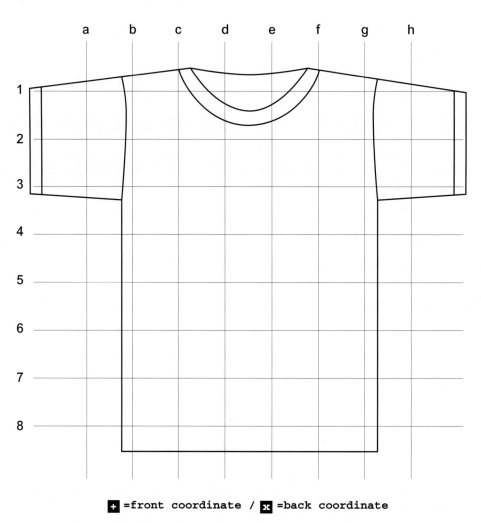

Figure 6.14 The form provided to workshop participants to help visualize how the coordinates will be placed on the t-shirt.

```
/*
* Example program: Code 1
*/
garment=(t-shirt); // this is a t-shirt program
wear(normal); // wear the shirt as you would normally
connectFront(g3,h3);(d5,b5);(g9,e9);(d3,a3); // connect these coordinates on
front
connectBack(d5,c5);(b3,a5);(b8,a5); // connect these coordinates on back
/*
* machine code (SMS-Tweet) of Code 1
*/
g=t-s,w=n;
cF(g3,h3/d5,b5/g9,e9/d3,a3);cB(d5,c5/b3,a5/b8,a5)
```

With this model Busch playfully presented his Fashing programming language to describe how notation for altering garments can be encoded. This is an excellent example of programmatic concepts applied to fashion. Even though the programming is not done on a computer and is hand-executed by the participants, the procedural logic adheres to rules of programming and the participants must understand the code in order to work successfully with the given constraints. This example suggests exciting possibilities for alternative design methods that can be applied to the traditional approaches with great success and can be distributed through digital means and social media.

INTERVIEW
OTTO VON BUSCH

Otto von Busch is a Swedish fashion artist, theorist, designer, crafter, and Haute Couture Heretic, as well as a DIY-demagogue. He aspires to liberate the mythical energy in fashion through hacking the flows of the operating system. For his PhD, *Fashion-able: Hacktivism and Engaged Fashion Design*,[3] von Busch explored how participatory practices can render fashion inclusive, yet still exclusive. To realize this he drew on the practice and tactics of hacking, fan fiction, liberation theology, and development practice through hands-on projects.

His research at the School of Design and Crafts at the University of Gothenburg in Sweden focused on "hacktivist" practices in design, fashion, and craft, identifying how such practices engage users to break the so-called "interpassivity" of consumerism.

He is a member of the National Swedish Handicraft Council and is engaged in the resurgence of educational *sloyd* (handicraft-based education) as a founding member of the Institute of Strategic Sloyd. Coming from an academic background, he is not a fashion designer by trade, but he has extensive knowledge in fashion design and has experience in sewing and craft techniques, and as a result has mixed fashion philosophy with fashion craftsmanship into a unique viewpoint.

AG: You are a craftsman, researcher and an activist. You completed your PhD, *Fashion-able: Hacktivism and Engaged Fashion Design*, at the School of Design and Crafts at the University of Gothenburg in 2008. How did you arrive at that topic?

OVB: I wrote my master's thesis about fashion theory while studying patternmaking and sewing in addition to material and interface design. That was a mix of interaction design and industrial design, so I gradually connected more and more with the idea of New Media and hacking and how this could come together with fashion design. If we imagine that fashion is an operating system, then I ask myself: "What is a new media approach to this operating system?" I am interested in the merger of the fashion system and the fashion logistics of material production with the myth production of fashion. How can a designer work with both of these "software approaches," how to hack into that system? Since I had the craft background, I felt that this was a hacker movement I was interested in. You go into the process and manipulate or "hack" the system, just like in craft you can manipulate the materials, rather than design them top down.

AG: Do you consider yourself a fashion designer or a hacktivist and disrupter who happens to work in the fashion domain?

OVB: It is always tricky when you have a mixed background. Are you a fashion designer if you are not established in the system? Are you an artist if you are not presenting in a gallery? To me I am definitely a fashion designer, but I work more to re-design how fashion designers work with the system. Rather than being a fashion designer who produces clothing, which is on the hanger in a store, I try to redesign or imagine and work with other ways to be a designer and work in relation to the fashion system.

AG: You started working on this approach years ago, but it seems that the rest of the design society

---

3. Otto von Busch, *Fashion-able: Hacktivism and Engaged Fashion Design*, PhD dissertation, accessed August 2, 2015, http://www.hdk.gu.se/sites/default/files/media/fashion-able_webanspassahd%20avhandling_OttovonBusch.pdf

is catching up with your ideas now, a few years after you.

OVB: It made sense to me at the point when I was making this progress; now it simply makes sense to more people.

AG: You have been using recycling as an ideology well before sustainability was in trend. Do you consider yourself a sustainability designer, or is this desire to "save the dying garments" rooted in a different philosophy?

OVB: Yes, I never really put any emphasis on sustainability. I put more emphasis on social sustainability. So much of sustainability, especially when I started this 10 years ago, was about materials—organic cotton, working conditions in factories—but I felt an important part of my work was asking what skills we acquire. How do they change our social relationships here and within society? How can we work with our feelings, skills, rather than look at the material contents and material production of clothes. It's much more about how do we produce a sustainable user, sustainable consumer, rather than sustainable production, which has been the main narrative in sustainability. And of course that is really important, but I look at sustainable practices that people do, rather than sustainable production. I have been looking at skills and craft interventions as a sphere I can work in. And these days the concept of sustainability is growing in this direction.

AG: Was there any resistance to your ideology and concepts?

OVB: The main resistance has come from designers themselves. They are the ones asking: "What are these amateurs doing? Why are my design skills not worth it, and why wouldn't people pay for my professional skills, and buy my garments?" I think the more you learn about the process, the more you appreciate the workmanship and designer clothes. The more trained and skilled I am, the more I appreciate the workmanship of professionals.

AG: You've been described as a hacker and a subversive, and much of your work in fashion is about questioning the system. How does that position translate into teaching fashion-related courses? How do you fit in a traditional fashion-design-driven educational system?

OVB: I am mostly "labeled" as an alternative and most of the teaching I do is about other values. But it is still about making garments. I teach about the role of the garment and how that can be different. It's about garments in an expanded sense in time and social engagement. I don't want to teach fashion designers to be consultants. I want them to be something more, to know the value of their skills and to consider garment production in an expanded sense in other realms.

AG: Do you feel that your students need to be at a certain level of craftsmanship in order to transition to hacktivism, or even to take your courses?

OVB: Students without skills sometimes have an expanded mind. Since they don't know the processes yet, they can be more creative. The main problem is that society is so saturated by fashion designers. Fashion is everywhere: on TV, in daily publications and news outlets. It is such a common phenomenon that students come preprogrammed knowing what fashion already is, and that is the tricky part. Fashion does not have to be exactly like the garments you see in the store. There are other types of fashion, and the tricky part is teaching that fashion does not have to be glamorous. There are very few designers who want to challenge what works. Most people do not challenge the system, and the whole system is not sustainable. It is built on a social and environmental unsustainability.

AG: In your foreword to the Artist Clothing catalogue 2004/2005, you say: "The fashion system is a metaphysic system closely guarded by sects and hermeneutic schools. These are all ruled by a small elite guard of professional mystics and interpreters—a leading group of cardinals and priests, following the directions and rules of a higher ideal." Now almost ten years later, with fashion collections presented online to a global audience and fashion bloggers sitting front row right next to the elite magazine editors, do you feel that the fashion system is finally open, or is it still ruled by an elite of select few?

OVB: The open system is an illusion. There are bloggers who have acquired a huge following and become famous, but who is inviting them? There is still the power of the one inviting them! There is an asymmetry in this process. As soon as a famous blogger loses their following, they won't be invited any more. Part of this democratization is an illusion. Very few of them are knowledgeable and educated on the craft of fashion and design. I think that this democratization has actually backfired. If you look at LookBook, there were so many different looks, but the popular ones are the ones that update their looks every day, so they are either wealthy or they have a connection to the stores and designers. It turns people into small magazines. So not everyone can be found; it's the people who have invested in the media scene, not necessarily invested in building a character. It's about building an Internet popular persona. That feeds into the capitalist traditional fashion system.

AG: In your workshop Fashion Fianchettos, draped maneuvers and fashion functions participants explored how fashion could be a set of mathematical functions, minimal codes of new drapings, which could even be sent between fashionistas as secret codes. Do you suggest that any part of the traditional fashion design process can become a set of mathematical rules, and how would that help the creative process?

OVB: The process for this workshop came out of frustration with fashion and technology being discussed mainly as technology integrated into the garment. I disagree with the idea that fashion and technology is just a combination of LEDs and wires, and I felt that there have to be other ways to look at technology outside the garment. The garment does not have to be a smart garment, but it can have a smart relationship with technology. Draping exercises can be creative exercises in coding the process. You can write down the steps or "coordinates" of how you draped a garment and then you can tweet or Facebook that code to other people, so they can program their own garment. A user could wear an oversized t-shirt, and there would be coordinates on the t-shirt of how to drape that t-shirt and maybe even just safety pin it. We all have the hardware at home, which is the t-shirt. Now we can receive the "software," which is easily followed and can be executed by anyone.

What I really tried to push in this workshop is the relationship of a technological societal system and "non-smart" garments and the opportunity to look at the technology outside the garment. What is interesting to me is a garment that is technologically enhanced and amplified by a technological system. This garment is not in production and does not contain smart fibers. You may even say it is a stupid garment, but it has a smart connection with social media, and that is what's fascinating me! To me this is the expanded sense of fashion and technology. The garment reaches into the user and the global system. I want us to become smarter about our relationship with the garments, instead of only focusing on garments with sensors.

For me there is a problem with the type of fashion and technology that responds to sensors measuring inner heartbeat or outside temperatures. It has a market but it brings out the worst of the Internet. It brings out surveillance, control, etc. Is this really what we want? I buy fashion to transform. Fashion has a higher mythical purpose for me. It has magic. And I was really frustrated. How do we raise our thinking? What is the magic of fashion and how do we reach that magic with technology?

AG: In your paper "Zen and the Abstract Machine of Knitting,"[4] you draw a parallel between "Knitting and Protocols," arguing that "even though protocols have an everyday connotation to software, they are ubiquitously present. Are you satisfying the needs of the disrupter in you to hack the "Machine of Knitting," or is this a philosophy of creative process combined with new media that any designer should look into?

OVB: This was more of a philosophy. I wanted to look at these micro skills! The making of a loop and purl. I'm not interested in the machines, but what are the basic skills we learn from them. How can we really rethink how we can use these micro skills not to reshape a silhouette, but how these micro skills connect to other micro skills. How protocols shape emergent behaviors, which connect to something else. It starts to mutate into something else and something totally different grows out.

And that's what's interesting about coming back to social media and draping garments. That t-shirt and the code we wrote about it can connect to larger system and just transform and mutate to something new and different. Rather than focus on the shape of a garment, that little micro behavior or protocol can actually shape that t-shirt and open a whole new world of thinking and applying fashion together with technology.

Fashion designers are thinking in an extremely limited sense and are stuck on the body and where to move the seams. If we apply these protocols and shift a little the way we connect practices, then totally different things can happen and that's what I wanted to pinpoint with this idea.

AG: When someone says "fashion and technology," what does that phrase mean to you?

OVB: Unfortunately, it most often evokes a frustration about a singular view of integrating smart circuits into a garment, which I support, because there are really smart people working in that area, but we are most often stuck in this techno fetishism about the garments themselves, and I would like to see it in a much more expanded sense. When people say "fashion and technology," I want us to think about programming in a larger sense too. Garments are programmed with certain behavior, certain rituals, and I'm thinking how do I reprogram this ritual and this use of the garment and my relationship with our value system of garments? I think that if we connect the smartness of garments and the garment itself with the other societal programs, that's when it becomes really interesting.

---

4. Otto von Busch. "Zen and the Abstract Machine of Knitting," (*Textile: The Journal of Cloth and Culture*, 11, no. 1, 2013: 6–19).

INTERVIEW
CAIT REAS

Cait Reas is an experienced fashion designer who is the founder of 1/1 studio. 1 of 1 is an independent design studio that synthesizes fashion and art into one-of-a-kind apparel that is made to order in Los Angeles. Each signed and numbered piece results from a collaboration between Cait Reas and a commissioned artist. She often collaborates with her husband Casey Reas, who writes software to explore conditional systems as art. His generative software becomes the core medium, which gets translated in various areas of visual experiences, including prints, objects, installations, and performances. One such collaboration between the two is shown in Figure 6.15 a, b, and c.

AG, KM: What is your background and how did you get involved in wearable technology and fashion?

CR: My background is in apparel design. I have a degree in apparel design and worked within the fashion industry for many years as a sportswear designer. I became interested in code and fashion through the work of my husband, Casey Reas. He is an artist who uses software to create images that are moving and evolving. I wanted to capture that movement by working his images into clothing and have the person who is wearing the

Figure 6.15 a, b, c The Yes No prints (left) by Casey Reas explore different expressions of the Yes No instructions, ranging from the analog display signal of a Commodore 64, to an emulator of the same platform, to new interpretations written for contemporary machines. The two dresses are designed by Cait Reas to implement the artwork as textile prints.

garment bring the movement back to life. My collaboration with Casey began in 2007 and is very different from the commercial work I was doing before. The work I was doing was often based on predicting trends and implementing them for a specific company's brand. The work that comes from our collaboration uses his images as my primary source of inspiration: they influence the color, fabrication, architecture, and silhouette of the garment.

AG, KM: What is your design process? How do you develop your ideas?

CR: My design process begins with the images Casey creates and building a concept from those images. For me it's important to develop a core idea behind the line and to design into those parameters. I develop ideas through drawing sketches, fabric research, and making hand swatches to try various techniques such as embroidery, beading, cut work. I'm very interested in creating texture and dimension through the surface treatment of the fabric. I like to make croqui with scaled versions of the actual images or treatments in Illustrator and Photoshop. If I'm having an image digitally printed, this is especially helpful to determine what scale to print the image and where it will be placed on a garment.

AG, KM: Many fashion designers have expressed dissatisfaction with computational and interactive technology. What are the limitations of the technology you are currently using in your designs and what would you like to see changed?

CR: With my work, any limitations relate to the cost of using specialized equipment and small production runs. I use local vendors to have fabric digitally printed, embroidered, and samples made, but because the work is all custom ordered, my production volume is very low and the cost of having the garments made is very high.

AG, KM: Your work is focused on producing single garments that are not offered at a commercial retail scale. Does working with software and code enable or support your production model? If so, how?

CR: The ability to create garments that are completely unique from one another is at the core of the studio. Each piece is signed and numbered and results from my collaboration with Casey. Still images seen on or within the garments are taken from moving images created by Casey's code and software. I am interested in responding directly to the form and quality of his art when creating new garments.

AG, KM: Tell us about your current work? Are you working with any new technologies, approaches, or techniques? How does that impact the project?

CR: My current work focuses on deep research into ways of translating Casey's art, his software images, into fabric and wearable garments. My first work began with digital printing his images and I still believe there are endless possibilities for the prints. However, I am also very interested in building texture and dimension through surface treatment of the fabric and how the images can be interpreted in a variety of ways.

AG, KM: What is your advice for future designers working in the field of fashion and technology? What are the opportunities and challenges facing them?

CR: The opportunity is that the space where fashion and technology meet is still relatively new and open. There is room to grow, define, and innovate. There are plenty of clothes out there; it's exciting to create something truly original or innovative. The challenge is creating something that can sustain continued research and creation. Can your ideas and products be produced and sold in order to facilitate the cycle of creating new ideas and products?

For Review

1. How do you define code?
2. Who is considered to be the first programmer and why?
3. What is a development environment?
4. What are some of the ways you can incorporate code into the design of garments and accessories?
5. How does the Kinematics system, created by Jessica Rosenkrantz and Jesse Louis-Rosenberg, create flexible structures?
6. What are the three most common characteristics of a programming language?
7. What is conditional structure?
8. How do you define looping structure?
9. Who created the Processing language?

For Discussion

1. How can an instruction set be utilized in the design process?
2. Into what kind of design would you integrate the Kinematics system?

Books for Further Reading

Bohnacker, Hartmut, Benedikt Gross, Julia Laub, and Claudius Lazzeroni. *Generative Design: Visualize, Program, and Create with Processing.* New York: Princeton Architectural, 2012.

Maeda, John, and Red Burns. *Creative Code.* Paris: Thames & Hudson, 2004.

Reas, Casey, and Ben Fry. *Processing: A Programming Handbook for Visual Designers and Artists.* Cambridge, MA: MIT, 2007.

Reas, Casey, Chandler McWilliams, and Jeroen Barendse. *Form Code in Design, Art, and Architecture.* New York: Princeton Architectural, 2010.

Shiffman, Daniel. *Learning Processing: A Beginners Guide to Programming Images, Animation, and Interaction.* Amsterdam: Morgan Kaufmann, Elsevier, 2009.

Key Terms

Bug: A term used by programmers to name an error in the code.

Cinder: A powerful toolbox for programming audio, video, graphics, image processing, and computational geometry.

Code: A system of rules that converts information into another form or representation for communication through a channel or for storage in a medium.

Conditional structures: Features of programming languages that perform different actions depending on whether a programmer-specified condition is evaluated as true or false; also known as conditional statements.

Creative coding: A growing practice and community of artists and designers who use programming to create expressive artifacts.

Development environment: A collection of procedures and tools for developing, testing, and debugging an application or program.

Looping structures: Processes that repeat themselves in a sequence of steps until a certain condition is true or as long as a certain condition remains true.

OpenFrameworks: OpenFrameworks v0.01 was released by Zachary Lieberman on August 3, 2005, and is an open source toolkit designed for creative coding.

Processing: Program created to serve as a software sketchbook and to teach computer-programming fundamentals within a visual context. It is a programming language, development environment, and an online community.

GLOSSARY

AC and DC current: AC means that the current is alternating, and DC means that the current is direct.

Additive fabrication: A fabrication process that creates an object through adding small amounts of a material until an object emerges.

Additive manufacturing (AM): Processes that sequentially deposit material in layers until the entire object is created.

Anode: The negative terminal in the cell of a battery.

Borgs: Early adopters, who were mostly computer science researchers, for whom the benefits of having a mobile computer at one's fingertips outweighed the drawbacks of its socially unacceptable appearance.

Bug: A term used by programmers to name an error in the code.

Cathode: The positive terminal in the cell of a battery.

Cinder: A powerful toolbox for programming audio, video, graphics, image processing, and computational geometry.

Circuit: A complete and closed pathway or a never-ending loop through which an electric current can flow uninterrupted.

CNC routers: A computer-controlled machine for cutting various hard and soft materials, such as wood, composites, aluminum, steel, plastics, foams, fabrics, leather, and other materials.

Code: A system of rules that converts information into another form or representation for communication through a channel or for storage in a medium.

Conditional structures: Features of programming languages that perform different actions depending on whether a programmer-specified condition is evaluated as true or false; also known as conditional statements.

Conductive fabrics: Woven or knit fabrics that vary in structure and carry an electrical curent.

Conductive loop fastener: A hook and loop fastener made from materials that allow it to conduct electricity.

Conductive materials: Materials with the ability to conduct or transmit heat, electricity, or sound; examples include conductive fabrics, threads, paints, and tapes.

Conductive paints: Compounds that contain copper, carbon, or silver; can be brushed or drawn onto fabric; and can conduct electricity.

Conductive tape: Usually made out of copper or aluminum, this tape can carry a current of electricity and act as a wire. It comes in various widths.

Conductive thread: Thread that carries a current of charged electrons just like a wire does; it can be spun from plated silver or stainless steel.

Conductive wool: Very fine steel conductive filaments, mixed with natural wool or polyester fibers, that can carry a current of electricity.

Conductor: An object, material or fabric that permits the flow of electric charges in one or more directions.

Continuity mode: This mode of a multimeter tests the resistance between two points and tells us if the two points are connected electrically. If they are, then the multimeter emits a tone. This test helps ensure that connections are made correctly between two points.

Creative coding: A growing practice and community of artists and designers who use programming to create expressive artifacts.

Current: The amount of electrical energy passing through a particular point, rated in Amps (A) or Milliamps (mA).

Development environment: A collection of procedures and tools for developing, testing, and debugging an application or program.

Digital fabrication: A fabrication process in which a machine is controlled by a computer.

Diode mode: A diode is an electrical component that only allows for electricity to flow in one direction. This test will show the proper direction in which to connect an LED. It is most often used in soft circuits to test if an LED is oriented correctly if there are no markings or if the + and − symbols cannot be seen.

DIY: Commonly used short term for "Do It Yourself." It defines the method of building, modifying, or repairing something by yourself, without the aid of experts or professionals.

Electrical resistance: An electrical conductor is the opposition or resistance to the flow of an electric current through that conductor.

Electric current: A flow of electric charge, traveling from a point of high electrical potential, identified as power or the symbol for plus (+), to the lowest, which is usually identified as ground or the symbol minus (-).

E-textile toolkit: Typically a kit that contains all the components and fabrics to create a simple soft circuit in order to make the acquisition of these components easier.

Extrusion: A process used to create objects of a fixed cross-sectional profile in which a material is pushed or pulled through a die of the desired cross-section. In this book, 3D printing refers to extrusion-type printing: raw material being melted and formed into a continuous 3D shape as defined by the design in the input file.

Fabricated kits: Narrower in scope than starter kits, these typically include only enough products to complete a set activity.

Fixed resistor: A component that adds resistance to a circuit at a predetermined value that cannot be changed.

Flexible circuit: A vast array of conductors bonded to a thin dielectric film.

Hertz (Hz): The unit of measure for frequency named for Heinrich Rudolf Hertz, who was the first to conclusively prove the existence of electromagnetic waves.

Hydrochromic inks: Inks that react as a result when exposed to water. They are usually transparent or invisible when dry and transform into a vibrant color when exposed to water.

Laser cutting: This process uses lasers in a subtractive digital fabrication process.

LED: A light-emitting diode; a special type of diode that lights up when electricity passes through it.

Looping structures: Processes that repeat themselves in a sequence of steps until a certain condition is true or as long as a certain condition remains true.

Maintained switches: Switches that retain their state until they are actuated into a new one.

Maker community: The group of people around the world who are becoming influenced to be DIY makers or are actively participating in the maker movement.

Maker movement: A contemporary culture or subculture representing a technology-based extension of DIY culture. Typical interests enjoyed by the maker culture include engineering-oriented pursuits, such as electronics, robotics, 3D printing, and the use of CNC tools, as well as more traditional activities such as metalworking, woodworking, and traditional arts and crafts.

Makerspace: A physical community space where people who consider themselves makers can get together and share resources, such as digital fabrication tools, electronics, and knowledge, to collaborate with each other.

Microcontroller: A small computer on a single integrated circuit that includes a processor core, memory, and programmable input/output peripherals (as defined by Syed R. Rizvi in *Microcontroller Programming*).

Momentary switches: Switches that stay in a particular state only as long as they are activated and return to their original state when released.

Ohm (Ω): The international unit of electrical resistance

OpenFrameworks: OpenFrameworks v0.01 was released by Zachary Lieberman on August 3, 2005, and is an open source toolkit designed for creative coding.

Photochromic materials: Change color with changes in light intensity or brightness. The normal state of the material is usually colorless or translucent, and it changes with exposure to a particular light source.

Photoresistor: A variable resistor controlled by light; also called a light-dependent resistor (shortened to LDR) a CdS cell (if it is made from Cadmium Sulfate), or a photocell.

Potentiometer: A variable resistor with two or three terminals and a sliding or otherwise movable contact element, called a wiper, which forms an adjustable voltage.

Processing: Program created to serve as a software sketchbook and to teach computer programming fundamentals within a visual context. It is a programming language, development environment, and an online community.

Reactive materials: Materials that respond to UV light, temperature changes, contact with water, and other external triggers.

Resistor: A component that adds resistance to a circuit and reduces the flow of electrical current.

Sintering: The process of compacting and forming a solid mass by heating a material without melting it to the point of liquefaction and depositing it at points in space defined by a pre-designed 3D model, binding the material together to create a solid structure.

SMD: Surface-mount device; any electronic device that is made through a method in which the components are mounted or placed directly onto the surface of printed circuit boards.

Starter kits: A typical starter kit will include electronic components (resistors, LEDs, pushbuttons, potentiometer), a prototyping board (breadboard), jumper wires, battery clip, USB cable, and a microcontroller development board (Arduino).

Subtractive fabrication: A fabrication process in which machines chisel or cut away at a block or sheet of material.

Switch: A component that creates a physical break in a circuit, which prevents the flow of electricity.

Thermochromic inks: Pigments or dyes that change color as a result of a temperature change.

UV color changing threads: Threads developed to respond to UV light and transform from no color to vibrant, bright colors or change from one color to another.

UV reactive paint and inks: Materials that change when exposed to UV/sunlight.

Variable resistor: A component that changes its values for reducing the flow of electrical current depending on outside influence.

Voltage: The difference in electrical energy between two points, rated in Volts (V).

Watt: The electrical unit of power, named after James Watt, a Scottish inventor who made improvements to the steam engine during the late 1700s and helped jumpstart the Industrial Revolution.

Wearables: Eyeglasses and the wristwatch are often cited as the first "wearables" in the history of wearable computing, as described by Bradley Rhodes on his website "A Brief History of Wearable Computing."

Resources

The following references are provided for further reading and to assist in sourcing materials for use with this book.

Books

FASHION AND TECHNOLOGY

Extreme Textiles: Designing for High Performance, Maltida McQuaid, Princeton Architectural Press, New York, 2005.
Fashionable Technology, Sabine Seymour, Springer Wien New York, New York, 2008.
Fashion Futures, Bradley Quinn, Merrell, London, 2012.
Fashioning the Future: Tomorrow's Wardrobe, Suzanne Lee, Warren du Preez, Thames & Hudson, United Kingdom, 2005.
Functional Aesthetics, Sabine Seymour, Springer-Verlag/Wien, New York, 2010.
Techno Fashion, Bradley Quinn, Berg/Bloomsbury Academic, London, 2002.
Techno Textiles: Revolutionary Fabrics for Fashion and Design, Sarah E. Braddock, Marie O'Mahony, Thames & Hudson, United Kingdom, 1999.
Techno Textiles 2: Revolutionary Fabrics for Fashion and Design, Sarah E. Braddock Clarke, Marie O'Mahony, Thames & Hudson, United Kingdom, 2008.
Textile Futures: Fashion, Design and Technology, Bradley Quinn, Bloomsbury Academic, London, 2010.
Textile Visionaries: Innovation and Sustainability in Textile Design, Bradley Quinn, Laurence King Publishing, London, 2013.

CODE AND PROCESSING

Creative Code: Aesthetics + Computation, John Maeda, Thames & Hudson, United Kingdom, 2004.

Generative Design: Visualize, Program, and Create with Processing, Hartmut Bohnacker, Benedikt Gross, Princeton Architectural Press, New York, 2012.

Learning Processing: A Beginner's Guide to Programming Images, Animation, and Interaction, Daniel Shiffman, Morgan Kaufmann, San Francisco, CA, 2008.

Make: Getting Started with Processing, Casey Reas, Ben Fry, O'Reilly Media, Sebastopol, CA, 2010.

The Nature of Code: Simulating Natural Systems with Processing, Daniel Shiffman, Magic Book Project, 2012.

Processing: A Programming Handbook for Visual Designers and Artists, Casey Reas, Ben Fry, MIT Press, Cambridge, MA, 2007.

PHYSICAL COMPUTING

Make: Getting Started with Sensors: Measure the World with Electronics, Arduino, and Raspberry Pi, Kimmo Karvinen, Tero Karvinen, Maker Media, San Francisco, CA, 2014.

Making Things Talk: Using Sensors, Networks, and Arduino to See, Hear, and Feel Your World, Tom Igoe, Maker Media, San Francisco, CA, 2011.

Physical Computing: Sensing and Controlling the Physical World with Computers, Dan O'Sullivan, Tom Igoe, Thomson Course Technology, Boston, MA, 2004.

CRAFT, FASHION, AND TECHNOLOGY DIY

Fashion Geek: Clothes Accessories Tech, Diana Eng, North Light Books, Cincinnati, OH, 2009.

Fashioning Technology: A DIY Intro to Smart Crafting, Syuzi Pakhchyan, Maker Media, San Francisco, CA, 2008.

Make: Getting Started with Adafruit FLORA: Making Wearables with an Arduino-Compatible Electronics Platform, Becky Stern, Tyler Cooper, Maker Media, San Francisco, CA, 2015.

Make: Wearable Electronics: Design, Prototype, and Wear Your Own Interactive Garments, Kate Hartman, Maker Media, Sebastopol, CA, 2014.

Switch Craft: Battery-Powered Crafts to Make and Sew, Alison Lewis, Potter Craft, New York, 2008.

Materials

The following companies sell many of the electronic components and materials mentioned in this book. Sparkfun and Adafruit are both excellent sources to locate LEDs, small quantities of conductive thread and fabric, batteries, and more.

ELECTRONICS

Sparkfun: http://www.sparkfun.com
Adafruit: http://www.adafruit.com

CONDUCTIVE FABRICS AND REACTIVE MATERIALS

Many conductive fabrics can be found from the electronics vendors listed below. Information about how to purchase and use other reactive materials is also included.

Conductive Fabrics
LessEMF: http://www.lessemf.com
Plug and Wear: http://www.plugandwear.com

Thermochromic Inks
XS Labs: http://www.xslabs.net/color-change/

UV Color Change Materials (also sells hydrochromic ink)
Solar Active: http://www.solaractiveintl.com

DIGITAL FABRICATION

Shapeways: http://www.shapeways.com
MakerBot: http://www.makerbot.com
Materialise: http://www.materialise.com/
Stratasys: http://www.stratasys.com/

Online Resources

The following websites have a great deal of information about materials, techniques, and resources for fashion and technology.
Kobakant: http://www.kobakant.at
Open Materials: http://openmaterials.org
Instructables: http://www.instructables.com

CREDITS

Chapter 1

1.0 Collier Schorr for Opening Ceremony and Intel, p. 2.
1.1 Stephanie McNiel, p. 4.
1.2 Universal Images Group/Getty Images, p. 5.
1.3 JONATHAN NACKSTRAND/AFP/Getty Images, p. 6.
1.4 Photography: Forster Rohner AG, p. 9.
1.5 Common rights, p. 10.
1.6 Ringly, p. 11.
1.7 Intel and Opening Ceremony, p. 11.
1.8 photo: David Klugston, p. 12.
1.9 SINEAD LYNCH/AFP/Getty Images, p. 13.
1.10 Adafruit Industries, www.adafruit.com, p. 14.
1.11 Philippe Huguen/AFP/Getty Images, p. 16.
1.12 Photography: James Kachan, p. 17.
1.13 Vega Edge, p. 18.
1.14 Vega Edge, p. 18.
1.15 Photography: James Kachan, p. 19.
1.16 Vega Edge, p. 20.
1.17 Joanna Berzowska, XS Labs © 2012 Photography: Ronald Borshan © 2012, p. 23.

Chapter 2

2.0 FRANCOIS GUILLOT/AFP/Getty Images, p. 28.
2.1 Stephanie McNiel, p. 32.
2.2a Aneta Genova, p. 32.
2.2b Stephanie McNiel, p. 33.
2.2c Stephanie McNiel, p. 33.
2.3 Paola Guimerans, p. 34.
2.4 Stephanie McNiel, p. 35.
2.5 Shawn Hempel/Shutterstock, p. 35.
2.6 Stephanie McNiel, p. 36.
2.7 Stephanie McNiel, p. 37.
2.8 Stephanie McNiel, p. 38.
2.9a Adafruit Industries, www.adafruit.com, p. 39.
2.9b Adafruit Industries, www.adafruit.com, p. 39.
2.10a Adafruit Industries, www.adafruit.com, p. 40.
2.10b Adafruit Industries, www.adafruit.com, p. 40.
2.11 KOBAKANT (Mika Satomi and Hannah Perner-Wilson), p. 41.
2.12 Paola Guimerans, p. 41.
2.13 Stephanie McNiel, p. 42.
2.14 Stephanie McNiel, p. 43.
2.15 Stephanie McNiel, p. 44.
2.16 Stephanie McNiel, p. 44.
2.17 Flip Switch by Katharina Bredies with Hannah Perner Wilson and Sara Diaz Rodriguez, p. 45.
2.18 Flip Switch by Katharina Bredies with Hannah Perner Wilson and Sara Diaz Rodriguez, p. 46.
2.19 Flip Switch by Katharina Bredies with Hannah Perner Wilson and Sara Diaz Rodriguez, p. 46.
2.20a Flip Switch by Katharina Bredies with Hannah Perner Wilson and Sara Diaz Rodriguez, p. 47.

2.20b	Flip Switch by Katharina Bredies with Hannah Perner Wilson and Sara Diaz Rodriguez, p. 47.	2.48b	Stephanie McNiel, p. 75.
2.21	Flip Switch by Katharina Bredies with Hannah Perner Wilson and Sara Diaz Rodriguez, p. 48.	2.49	Stephanie McNiel, p. 76.
		2.50	KOBAKANT (Mika Satomi and Hannah Perner-Wilson), p. 77.

Chapter 3

2.22	Stephanie McNiel, p. 51.
2.23	Stephanie McNiel, p. 52.
2.24	Stephanie McNiel, p. 52.
2.25	Stephanie McNiel, p. 53.
2.26	Stephanie McNiel, p. 53.
2.27	Stephanie McNiel, p. 54.
2.28	Stephanie McNiel, p. 54.
2.29	Stephanie McNiel, p. 55.
2.30	Stephanie McNiel, p. 55.
2.31a	Stephanie McNiel, p. 57.
2.31b	Stephanie McNiel, p. 57.
2.32a	Stephanie McNiel, p. 58.
2.32b	Stephanie McNiel, p. 58.
2.33a	Stephanie McNiel, p. 59.
2.33b	Stephanie McNiel, p. 59.
2.34	Stephanie McNiel, p. 60.
2.35a	Stephanie McNiel, p. 62.
2.35b	Stephanie McNiel, p. 62.
2.36a	Stephanie McNiel, p. 63.
2.36b	Stephanie McNiel, p. 63.
2.37a	Stephanie McNiel, p. 64.
2.37b	Stephanie McNiel, p. 64.
2.38a	Stephanie McNiel, p. 65.
2.38b	Stephanie McNiel, p. 65.
2.39a	Stephanie McNiel, p. 66.
2.39b	Stephanie McNiel, p. 66.
2.40a	Stephanie McNiel, p. 67.
2.40b	Stephanie McNiel, p. 67.
2.41	Stephanie McNiel, p. 68.
2.42a	Stephanie McNiel, p. 70.
2.42b	Stephanie McNiel, p. 70.
2.43a	Stephanie McNiel, p. 71.
2.43b	Stephanie McNiel, p. 71.
2.44a	Stephanie McNiel, p. 72.
2.44b	Stephanie McNiel, p. 72.
2.45a	Stephanie McNiel, p. 73.
2.45b	Stephanie McNiel, p. 73.
2.46	Stephanie McNiel, p. 74.
2.47a	Stephanie McNiel, p. 74.
2.47b	Stephanie McNiel, p. 74.
2.48a	Stephanie McNiel, p. 75.
3.0	Aimee Winters, Rainbow Winters, www.rainbowwinters.com, p. 84.
3.1	Stephanie McNiel, p. 87.
3.2	Photo: David Klugston, p. 88.
3.3	Stephanie McNiel, p. 89.
3.4	Stephanie McNiel, p. 90.
3.5	Stephanie McNiel, p. 90.
3.6	Stephanie McNiel, p. 91.
3.7	Stephanie McNiel, p. 92.
3.8	Stephanie McNiel, p. 93.
3.9	Stephanie McNiel, p. 94.
3.10a	Stephanie McNiel, p. 96.
3.10b	Stephanie McNiel, p. 97.
3.11a	THEUNSEEN, photo by Jonny Lee, p. 98.
3.11b	THEUNSEEN, photo by Jonny Lee, p. 99.
3.12	Aimee Winters, Rainbow Winters, www.rainbowwinters.com, p. 100.
3.13	Photo: Aneta Genova, p. 100.
3.14	Adafruit Indsutries www.adafruit.com, p. 101.
3.15	Elise Co, p. 102.
3.16	Elise Co, p. 102.
3.17	Stephanie McNiel, p. 103.
3.18	Stephanie McNiel, p. 104.
3.19	Stephanie McNiel, p. 104.
3.20	Stephanie McNiel, p. 105.
3.21	Stephanie McNiel, p. 105.
3.22	Stephanie McNiel, p. 106.
3.23	Stephanie McNiel, p. 106.
3.24	Stephanie McNiel, p. 107.
3.25	Stephanie McNiel, p. 107.
3.26	Stephanie McNiel, p. 108.
3.27	Stephanie McNiel, p. 108.
3.28	Stephanie McNiel, p. 109.
3.29	Stephanie McNiel, p. 109.
3.30	Stephanie McNiel, p. 110.
3.31	Stephanie McNiel, p. 110.
3.32	Stephanie McNiel, p. 111.
3.33	Stephanie McNiel, p. 111.
3.34	Stephanie McNiel, p. 112.

3.35 Stephanie McNiel, p. 112.
3.36 Stephanie McNiel, p. 113.
3.37 Stephanie McNiel, p. 113.
3.38 Stephanie McNiel, p. 114.
3.39 Stephanie McNiel, p. 114.
3.40 Aneta Genova, p. 115.

Chapter 4

4.0 Concept and design by Lesia Trubat (www.lesiatrubat.com) Photo: Lucia Jarque, Dancer: Triana Botaya, p. 120.
4.1a Stephanie McNiel, p. 124.
4.1b Stephanie McNiel, p. 125.
4.2 Stephanie McNiel, p. 126.
4.3 Stephanie McNiel, p. 127.
4.4 Instructables, p. 128.
4.5 Concept and design by Lesia Trubat (www.lesiatrubat.com) Photo: Lucia Jarque, Dancer: Triana Botaya, p. 130.
4.6 Concept and design by Lesia Trubat (www.lesiatrubat.com) Photo: Lucia Jarque, Dancer: Triana Botaya, p. 130.
4.7 Concept and design by Lesia Trubat (www.lesiatrubat.com) Photo: Lucia Jarque, Dancer: Triana Botaya, p. 131.
4.8 Stephanie McNiel, p. 131.
4.9 Stephanie McNiel, p. 132.
4.10 Stephanie McNiel, p. 133.
4.11 Stephanie McNiel, p. 134.
4.12 Stephanie McNiel, p. 135.
4.13 Anni Norddahl, p. 137.
4.14 Anni Norddahl, p. 138.
4.15 Diffus Design studio, p. 139.
4.16 Diffus Design studio, p. 140.
4.17 Stephanie McNiel, p. 144.
4.18 Stephanie McNiel, p. 144.
4.19 Stephanie McNiel, p. 145.
4.20 Stephanie McNiel, p. 145.
4.21 Stephanie McNiel, p. 146.
4.22 Stephanie McNiel, p. 146.
4.23 Stephanie McNiel, p. 147.
4.24 Stephanie McNiel, p. 147.
4.25 Adafruit Industries www.adafruit.com, p. 148.

Chapter 5

5.0 Chelsea Lauren Getty Images, p. 154.
5.1a Photography: Stephanie McNiel, p. 157.
5.1b Andy Kropa/Getty Images, p. 157.
5.2 Stephanie McNiel, p. 159.
5.3 Stephanie McNiel, p. 160.
5.4 Stephanie McNiel, p. 160.
5.5 Aneta Genova, p. 161.
5.6 Aneta Genova, p. 161.
5.7 Vereshchagin Dmitry/Shutterstock, p. 162.
5.8a threeASFOUR, p. 163.
5.8b threeASFOUR, p. 163.
5.9 Stephanie McNiel, p. 164.
5.10 Rendering by Bradley Rothenberg for threeASFOUR, p. 165.
5.11 Rendering by Bradley Rothenberg for threeASFOUR, p. 166.
5.12 Rendering by Bradley Rothenberg for threeASFOUR, p. 167.
5.13 Rendering by Bradley Rothenberg for threeASFOUR, p. 167.
5.14 Andy Kropa/Getty Images, p. 168.
5.15 Aimee Kestenberg Collection/ www.AimeeKestenberg.com, p. 169.
5.16 TRIPTYCH NY, p. 170.
5.17 TRIPTYCH NY, p. 171.
5.18 TRIPTYCH NY, p. 172.
5.19a TRIPTYCH NY, p. 172.
5.19b TRIPTYCH NY, p. 172.
5.20 TRIPTYCH NY, p. 173.
5.21 Aneta Genova, p. 174.
5.22 Stephanie McNiel, p. 175.
5.23 Stephanie McNiel, p. 175.
5.24 Stephanie McNiel, p. 176.
5.25 Stephanie McNiel, p. 176.
5.26 Jay Padia, p. 177.
5.27 Stephanie McNiel, p. 178.
5.28a Design and rendering by Jay Padia, p. 178.
5.28b Design and rendering by Jay Padia, p. 178.
5.29a Design and rendering by Jay Padia, p. 179.
5.29b Design and rendering by Jay Padia, p. 179.
5.30 Jay Padia, p. 180.
5.31 Stephanie McNiel, p. 181.
5.32a Stephanie McNiel, p. 181.

5.32b	Stephanie McNiel, p. 181.	6.5	Nervous System http://nervo.us, p. 199.
5.33	Stephanie McNiel, p. 182.	6.6	The Processing Foundation, p. 203.
5.34	Stephanie McNiel, p. 182.	6.7	The Processing Foundation, p. 203.
5.35	Cindy Ord/Getty Images, p. 183.	6.8	The Processing Foundation, p. 204.
5.36	Stephanie McNiel, p. 187.	6.9	Created by Katherine Moriwaki on Processing.org, p. 205.

Chapter 6

6.0	Kinematics Dress by Nervous System http://nervo.us, p. 192.	6.10	The Processing Foundation, p. 205.
6.1	Image courtesy of the bitforms gallery, New York, p. 196.	6.11	Aneta Genova, p. 206.
		6.12	Otto Von Busch, p. 208.
		6.13	Otto Von Busch, p. 209.
		6.14	Otto Von Busch, p. 210.
6.2	Jeremy Rotsztain 2011, p. 197.	6.15a	Yes No, 2013 by Casey Reas, p. 216.
6.3	LIA www.liaworks.com, p. 197.	6.15b	Dress by Cait Reas, p. 216.
6.4	Stephanie McNiel, p. 198.	6.15c	Dress by Cait Reas, p. 216.

INDEX

3D modeling
    bracelet, 177–178, 180
    overview, 156, 162
    shoes, 166

3D printing
    additive manufacturing, 162
    as fabrication technique, 22, 155–156, 199
    Bradley Rothenberg, 187–188
    design aesthetics, 15
    DIY and Maker Movement, 10, 27, 221
    extrusion, 191, 220
    for artists, 190
    threeASFOUR, 166, 183, 186
    tutorial, 177

A
AC current, 30
Adafruit Industries
    Becky Stern, 148–150
    DIY electronics, 129
    electroluminescent (EL) panel, 101
    Maker community, 14
    potentiometers, 39–40
Additive fabrication, 158, 190–191, 219
Additive manufacturing (AM), 162, 191, 219
Anode, 37, 40, 82, 219
Asfour, Gabi, 183

B
Bare Conductive, 91, 92, 115–116
Berzowska, Joanna, 21, 23, 227
Borgs, 9, 10, 27, 219
Bredies, Katharina, 29, 45, 49, 227
Bug, 80, 200, 218–220

C
Cathode, 37, 40, 82, 219
Cinder, 195, 201–202, 218–219
Circuit
    Bare Conductive paint, 92, 115–116
    basic components, 36–44
    CircuitWorks ®, 91
    CuteCircuit, 8

Climate Dress, 138, 142
components, 36–43, 152
connectivity test, 143–147
electric, 31–32, 91
electronic, 89, 123
flexible, 34–35
flip switch, 45, 49
for review and discussion, 81, 118, 151
Forster Rohner, 142
innovations, 149
Kobakant, 77, 79, 80
microcontroller, 130
overview, 7, 29–33, 82, 219–221
peripherals, 133–136
series vs. parallel, 32, 35, 56–60
smart, 215
soft, 51–55, 86, 87, 93
Vega Edge, 21
with zipper switch, 61–68, 69–76
with hook and eye closure, 103–108,
with metal snap closure, 109–114
Climate Dress, 137–140
Closures
    fashion industry fasteners, 92
    for review and discussion, 118
    hook and eye, 93, 103–108
    metal snaps, 109–114
CNC routers, 158, 191, 219
Code, 8, 169, 193–218, 223–224
Computational fashion
    collaborative process, 136
    conductive thread, 86
    design aesthetics, 15
    DIY, 123,
    EL panel, 102
    electronics and toolkits, 122–123, 129
    resources, 127
Conditional structures, 200, 218, 219
Conductive hook/eye closure, 103
Conductive fabrics
    electrical shielding, 7
    for review and discussion, 118
    overview, 85, 89–90, 119, 219
    research and development, 13
    resources, 224
Conductive loop fastener, 89, 119, 219
Conductive materials
    electricity, 30
    fabrics, 90, 119
    for review, 118
    hook and loop fasteners, 90, 119
    Kobakant, 78,
    overview, 85–87, 219
    paint, 92, 119
    tape, 91, 119
    wool, 89, 119
Conductive paints, 91, 119, 219
Conductive tape, 36, 46, 91, 92, 118, 119, 219
Conductive thread
    battery holders, 134
    Climate Dress, 138
    continuity test, 143, 146, 147
    electric circuit, 32, 36
    embroidered sensors, 11–12
    flip switch, 45
    hook and eye switch, 93, 103–108
    Katharina Bredies, 49
    loop fastener, 89
    maintained zipper switch , 64, 66, 68, 69, 72, 74, 76
    Maker movement, 14
    metal snap switch, 43, 93, 109–114
    momentary zipper switch, 64, 66, 69, 72, 74, 75
    overview, 86, 119, 220
    resources, , 224
    soft circuit, 51, 52, 54–56, 58
    toolkits, 125, 127
    zipper switches, 44, 61, 62
Conductive wool, 89, 119, 220
Conductor, 31, 34, 50, 82, 93, 220
Continuity mode, 144, 152, 220
Creative coding, 195, 200, 201, 218, 220, 221

D

DC current, 30, 82, 219
Design aesthetics, 15–27
Development environment, 129, 195, 200, 201, 218, 220, 221
Diffus Design Group, 140
Diffus Studio, 137
Digital fabrication, 5, 8, 15, 152, 155–191
Digital Multimeter Connectivity Test, 143
Digital Multimeter Continuity Test, 143
Diode mode, 145, 152, 220
DIY
    Adafruit, 148–150
    books, 81, 151, 224
    Climate Dress, 136–139
    electronics, 121, 123, 128, 129, 148, 151
    for review and discussion, 27, 151
    Maker community, 10, 27, 79, 133, 163, 220
    Maker culture, 15
    Maker movement, 13, 27
    overview, 3, 22
    Otto Von Busch, 224
    toolkits, 20, 121–152, 157
DIY electronics, 122–129, 148, 151
DIY Kit, 20, 121–152

E

Electrical resistance, 31, 82, 87, 220, 221
Electricity
    basics, 29, 30–32, 34
    conductive materials, 85, 86, 89, 93, 119, 219
    continuity test, 143–147, 152, 221
    flip switch, 45
    flow through circuit components, 36, 39, 40, 42, 82, 220
    for review and discussion, 81
    Kobakant, 78–79,
    soft circuit, 51
    zipper switch, 62
Electric circuit, 29–33, 35, 51

Electric current, 30, 31, 80, 82, 86, 219, 220
Electronic fashion, 3, 8–15, 149
Electronics
    Adafruit, 148–150
    Arduino, 14
    books, 81, 151, 224
    Climate Dress, 139, 142
    community, 121, 123, 157–158, 221
    components, 36–42
    DIY, 128–129
    flexible circuits, 32–34, 91, 115–117
    for review and discussion, 27, 151
    integration in garments, 29, 30, 86
    Katharina Bredies, 49–50
    Kobakant, 77–80
    LilyPad Arduino, 131
    overview, 5, 122–123
    peripherals, 133
    resources, 127
    soft circuit tutorials 51–55, 56–60
    toolbox, 123–126
    Vega edge, 17, 20, 21, 24, 25
    wearable, 8, 13, 148
    zipper switch tutorials, 61–68, 69–76
E-Textile
    crafts, 78–80
    DIY resources online, 128–129
    Forster Rohner, 139
    Katharina Bredies, 49–50
    overview, 122, 152, 220
    sensor, 88
    toolkit, 122, 129–135, 149
Extrusion, 95, 162, 191, 220

F

Fabricated kits, 123, 152, 220
Fashion Fianchettos, 208, 214
Fixed resistor, 37, 81, 82, 220
Flexible circuit, 34, 81, 82, 220

Flexible electronics, 34, 81, 82, 220
Flip switch, 29, 45, 46, 49, 227

G
Guglielmi, Michel, 140

H
Hartman, Kate:, 17, 20, 81, 151, 224
Hertz (Hz), 30, 82, 220
Hydrochromic inks, 85, 95, 119, 220

I
In series
    batteries, 37
    circuit, 30, 32, 34, 56, 60
    for review and discussion, 81
    LEDs, 35, 56, 60
    resistors, 42
Invisible stitch line, 51

K
Kobakant, 41, 77, 78, 81, 118, 225, 227, 228

L
Laser cutting
    digital fabrication overview, 155–156, 158–159
    future trends, 169
    overview, 10, 15, 191, 220
    threeASFOUR, 183, 186
    tutorial, 174–176
    Vega edge, 21–22
LEDs
    circuit components 36
    Climate Dress, 137–142
    connectivity test, 143, 147

flip switch, 45–48
hook and eye closure tutorial, 103–108
in parallel 34, 35, 56–59, 115
in series, 35, 56, 60
kits, 123, 125, 127, 152, 221
metals snap closure tutorial, 109–114
Otto Von Busch, 214
overview, 39
resources, 224
sewable, 42, 133, 136, 137–140
soft circuit tutorial, 51–55
zipper switch tutorials, 61–68, 69–76
Lizardi, Isabel, 115, 116
Loop fasteners, 89
Looping structures, 200, 218, 220

M
Maintained switch, 43, 44, 61–66, 68–74, 76, 81, 83, 220
Maker community, 10, 14, 15, 27, 220
Maker movement, 3, 13, 14, 27, 157, 220
Makerspace, 136, 152, 221
Metal fasteners, 92, 93
Microcontroller
    Arduino, 129, 131, 132, 133
    books, 151
    Climate Dress, 138, 149
    collaborative process, 136
    Electronic Traces, 130
    e-textile toolkit, 129
    expanded kit, 125
    for review and discussion, 151
    overview, 10, 121, 123, 152, 152
Momentary switch, 43, 44, 61–67, 69–75, 81, 82, 93, 136, 221

O
Ohm (Ω), 31, 37, 39, 42, 82, 143, 221
openFrameworks, 195, 202, 218, 221

P

Parallel
- batteries, 37
- for review and discussion, 81
- LEDs, 34, 35, 42, 56, 58, 59, 103, 109, 115
- Otto Von Busch, 215
- Overview, 30, 32

Peripherals
- LilyPad, 133, 135, 136
- overview, 121, 130, 132, 152, 221

Perner-Wilson, Hannah, 77, 81, 118, 227, 228

Photochromic materials, 95, 119, 221

Photoresistor, 38, 82, 221

Plastic color changing resin, 95

Popularizing tools, 14

Potentiometer
- circuit components, 36
- embroidered, 41
- expanded toolbox, 123
- linear, 39
- resistor, 38, 82
- soft, 40
- starter kits, 152, 221

Processing
- books, 218, 223
- development environments, 195
- for review and discussion, 218
- introduction to code, 193
- programing language, 200–201, 218
- learning code tutorial, 202–207

R

Reactive materials
- for review and discussion, 118
- hydrochromic inks, 95
- overview, 85, 86, 119, 221
- photochromic materials, 95
- plastic color changing concentrates, 95
- resources, 224
- thermochromic inks, 94, 96, 97
- UV color changing thread, 95, 100
- UV reactive paints and inks, 95, 100
- wind reactive inks, 95, 98, 99

Reas, Cait, 196, 216, 229

Resistor
- circuit components, 36–41,
- expanded toolbox, 123
- fixed, 37
- photoresistor, 38
- potentiometer, 38
- starter kits, 152, 221
- variable, 37, 38, 89

Rothenberg, Bradley
- 3D printed interlocking shapes, 163, 198
- 3D modeled shoes, 165–167
- interview, 187

S

Satomi, Mika, 77, 81, 118, 227, 228

Sensors
- Arduino, 129
- books, 224
- capacitive, 116
- embroidered, 11
- Flora, 129, 149
- for review and discussion, 151
- garments, 214
- handcrafted, 78, 79, 101–102, 129
- LilyPad, 130, 132, 133, 135, 136
- ribbon, 40
- textile, 24, 25, 49, 88
- switch, 42

Sewable LEDs, 42, 125, 127, 133, 136

Sew-on snaps, 93, 94

Sintering, 162, 187, 191, 221

Soft circuit
- bare conductive electric paint, 115
- Climate Dress, 138, 142
- conductive materials, 7, 86, 87, 89, 93
- e-textile toolkit, 220

flexible, 34, 35
for review and discussion, 81, 118, 151, 152
innovations, 149
Kobakant, 77–79
LilyPad peripherals, 133–136
overview, 29, 30, 33, 38
tutorial: conductive hook and eye closure, 103–18
tutorial: continuity test, 143, 146–147
tutorial: LEDs in parallel vs. series, 56–60
tutorial: metal snaps as closures, 109–114
tutorial: momentary and maintained switch with long zipper, 69–76
tutorial: momentary and maintained switch with short zipper, 61–68
tutorial: visible and invisible stitch line, 51–55
Starter kits, 115, 123, 152, 220, 221
Stern, Becky, 14, 148, 224
Subtractive fabrication, 158, 191, 221
Surface Mound Devices (SMD), 45, 82, 221
Switch
    books, 151, 224
    conductive loop fastener, 89
    flip switch, 29, 45–49
    for review and discussion, 81, 151
    hook and eye, 93, 103, 108
    LilyPad, 133, 134, 136
    maintained, 43, 44, 61–68, 69–76
    metal snaps, 43, 109, 114
    momentary, 43, 44, 61–68, 69–76
    overview, 30, 36, 42, 82, 86, 220–221
    photoresistor, 38
    POM POM ™ dimmer, 87
    potentiometer, 38
    zipper, 93, 61–68, 69–76

T

Thermochromic inks
    for review and discussion, 118
    on textiles, 96–97
    overview, 94, 119, 221
    resources, 224
threeASFOUR
    3D printed dress, 157
    3D printed interlocking shapes, 163
    Bradley Rothenberg collaboration, 165–168, 187
    interview, 183–186
    shoes, 165–168
Toolkit
    DIY, 157
    e-textile, 122, 129–135, 220, 221
    invent–abling, 126–127
    LilyPad Arduino, 142
    opensource, 218
    overview, 152
Treatments, 159, 196, 217
Triptych, 170, 171, 173

U

Ursomarzo, Tania, 170, 171
UV color changing threads, 95, 119, 221
UV color changing translucent beads, 100
UV reactive paint, 95, 119, 221
UV reactive ink, 100

V

Variable resistor, 37, 38, 82, 89, 221
Vega Edge, 17, 19, 20, 21, 22, 227
Visible stitch line, 51
Von Busch, Otto, 208, 212, 215, 229
Voltage
    batteries, 37
    electricity, 30, 31
    LEDs, 42
    multimeter, 143
    overview, 82, 221
    potentiometer, 38

W

Wearables
- Adafruit, 148–149
- books, 151, 224
- computers, 10
- embedded, 15
- for review and discussion, 27
- overview, 3, 8–9, 221
- Toronto Meetup, 21
- Katharina Bredies, 50
- LilyPad, 131

Wearable computers, 3, 9, 10, 11

Wearable electronics
- Adafruit, 14, 148
- books, 81, 151, 224
- Kate Hartman, 21
- research and design, 13